Th. Rüedi   A.H.C. von Hochstetter
R. Schlumpf

Surgical Approaches for Internal Fixation

Translated by T.C. Telger

Foreword by Martin Allgöwer

With 99 Figures, Partly in Color

Springer-Verlag
Berlin Heidelberg NewYork Tokyo 1984

Prof. Dr. med. THOMAS RÜEDI
Chirurgische Klinik, Rätisches Kantons- und Regionalspital Chur, CH-7000 Chur

Prof. Dr. ARTHUR H.C. VON HOCHSTETTER
Abteilung für topographische und klinische Anatomie, Departement für Chirurgie,
Kantonsspital, CH-4031 Basel
formerly o. Professor of Anatomy, University of Western Ontario, London,
Ontario, Canada

ROBERT SCHLUMPF
Löwengasse 176, CH-9620 Lichtensteig

Translator:

TERRY TELGER
3054 Vaughn Avenue, Marina, California 93933, USA

Translation of the German edition

Operative Zugänge der Osteosynthese
© Springer-Verlag Berlin Heidelberg 1984

ISBN-13: 978-3-642-69348-9   e-ISBN-13: 978-3-642-69346-5
DOI: 10.1007/978-3-642-69346-5

Typesetting, printing, and bookbinding: Universitätsdruckerei H. Stürtz AG, Würzburg.
2124 3130-543210.

Foreword

Surgeons confronted with acute trauma are frequently under great pressure to act quickly. Only a few have an infallible three-dimensional memory as regards the different approaches necessary for treating fractures by internal fixation. Thus there is a real need for a reference book on the approaches to the shoulder, arm, pelvis, and leg which is instructive and based on clinical practice. This is true both for the emergency situation and for the "evening before" with the imperative preoperative planning.

THOMAS RÜEDI, himself a surgeon as well as a gifted illustrator, in cooperation with ARTHUR VON HOCHSTETTER, a clinical anatomist, and excellently interpreted by the artist ROBERT SCHLUMPF, has created a novel and impressive atlas. The surgical approaches are depicted in a manner which is anatomically correct, limited to the essentials, and realistic. In addition, the attractive, black-and-white illustrations of the anatomy are successfully supplemented by color schematic drawings.

This luxuriously prepared edition may become a daily advisor to surgeons dealing with trauma. It deserves a widespread distribution in surgical departments and reference libraries.

Basel, Fall 1983   M. ALLGÖWER

Preface

Internal fixation has become an established part of fracture therapy. However, success with the method requires more than an accurate, stable reconstruction of the bone; it relies in large measure on the atraumatic handling of the surrounding soft tissues. Damage to nerves or blood vessels can cause atrophy and loss of function, while carelessly places incisions can leave disabling scars. Thus, the surgical approach to the bone is a matter of crucial importance in operative fracture management. It was this consideration that prompted us to conceive this book in close cooperation between surgeon, anatomist, and artist.

The atlas illustrates surgical approaches to the most common sites of bone and joint injuries. Each approach has been designed to give the best possible exposure with a minimum degree of trauma. Following the example of Henry's "extensile exposures," we have placed special emphasis on incisions which can be readily extended during the course of the operation, since the full scope of an injury often cannot be appreciated on preoperative films. Of the many approaches that have been devised previously, only one is presented for each surgical goal so that it may be more easily remembered. Each approach has been worked out in anatomical preparations, tested clinically, and illustrated by our artist.

Most of the skin incisions are linear so that they can be extended without creating flaps that are prone to edema and necrosis. Although a number of these incisions are made without regard for Langer's lines (at the knee, for example), the cosmetic results are still acceptable if a careful skin suture is applied. We feel that tendons, muscles, and ligaments generally should be osteotomized at their insertions, rather than divided. Later they can be reattached with screws or a tension band wire more simply and more securely than with sutures.

It will be noted that the illustrations are accompanied by a minimum of text. This reflects our feeling that the visual perception of anatomical and topographical relationships is a more powerful teaching tool than words alone.

We are indebted to our colleagues at the Division of Topographic and Clinical Anatomy especially Miss B. SCHERRER and Mr. O.S. BALDOMERO, Department of Surgery, University of Basel. We also thank Mrs. T. GOTSCH and Mrs. J. SOLIVA of the Surgical Clinic, Kantonsspital Chur, for typing the manuscript. In addition, we are grateful to Dr. GOETZE, and the employees of Springer Verlag for their patient support during preparation of atlas, and to Prof. M. ALLGÖWER for suggesting the concept of the book.

TH. RÜEDI
A.H.C. VON HOCHSTETTER
R. SCHLUMPF

Note About the Figures

Area of bone which may be exposed by the approach described in that chapter

Skinincision with its extension possibilities

Incision of fascia, muscles etc.

Osteotomy or detachment of insertions of tendons or ligaments

Contents

Part I   **The Upper Extremities**   1

1   Clavicle and Acromio-clavicular Joint   3
2   Scapula and Shoulder Joint: Posterior Approach   9
3   Proximal Humerus: Anterior Approach   15
4   Humeral Shaft: Antero-lateral Approach   21
5   Humeral Shaft: Posterior Approach   29
6   Distal Humerus and Humero-ulnar Joint   37
7   Humero-radial Joint and Radial Head   43
8   Elbow Joint and Proximal Radius: Volar Approach   49
9   Radial Shaft: Dorso-lateral Approach   57
10   Ulnar Shaft   63
11   Distal Radius: Volar Approach   69

Part II   **The Lower Extremities**   75

12   Sacro-iliac Joint and Posterior Iliac Crest   77
13   Symphysis Pubis and Superior Pubic Rami   85
14   Hip Joint: Posterior Approach   93
15   Hip Joint: Postero-lateral Approach   101
16   Hip Joint: Anterior Ilio-inguinal Approach   109
17   Proximal Femur: Lateral Approach   115
18   Femoral Shaft: Lateral Approach   123
19   Distal Femur: Lateral Approach   131
20   Patella   139
21   Knee Joint and Proximal Tibia: Lateral Parapatellar Approach   . .   143
22   Tibial Shaft: Anterior Approach   155
23   Tibial Shaft: Postero-medial Approach   159
24   Medial Malleolus and Distal Tibia   167
25   Lateral Malleolus   175
26   Calcaneus: Lateral Approach   181

References   185

Subject Index   187

The Upper Extremities

Part 1

The Upper Extremities

1 Clavicle and Acromio-clavicular Joint

Indications Acromio-clavicular dislocation, nonunion of the clavicle, fractures of the clavicle.

Positioning/ Draping Supine with ample padding beneath the shoulder (vacuum mattress). The arm is draped so that it is free to be moved in any direction. The head is turned toward the opposite side.

Skin Incision A sagittal incision about 5 cm long is made across the clavicle 1 fingerwidth medial to the *coracoid process*. The platysma and the lateral supraclavicular nerves are divided as the incision is made.

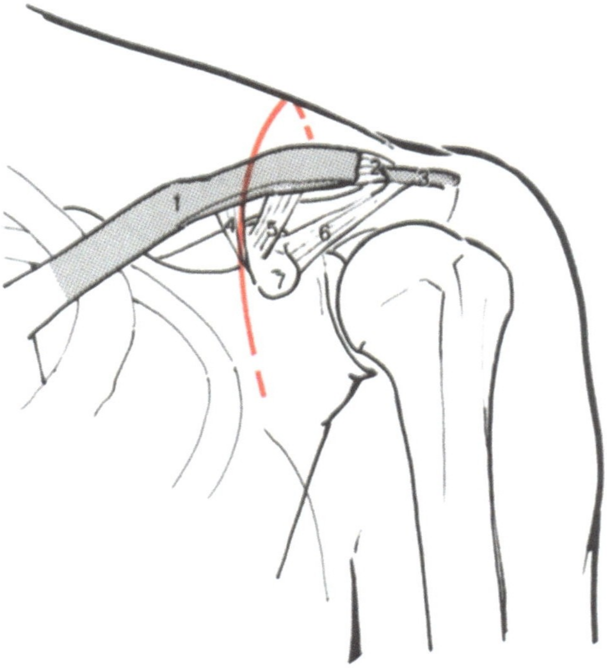

Fig. 1 a

1 Clavicle
2 Acromio-clavicular joint
3 Acromion
4 Conoidal ligament
5 Trapezoidal ligament
6 Coraco-acromial ligament
7 Coracoid process

1   Clavicle and Acromio-clavicular Joint

Deep Dissection   The anterior border of the clavicle is exposed medially and laterally. The deltoid muscle is partly detached from the clavicle; it may also be necessary to divide the clavicular attachement of the pectoralis major muscle, leaving a narrow strip of tissue on the bone.
Beware of injury to the cephalic vein!

Fig. 1 b

1   Intermediate and lateral supraclavicular nerves
2   Acromio-clavicular joint
3   Deltoid muscle
4   Pectoralis major muscle
5   Cephalic vein

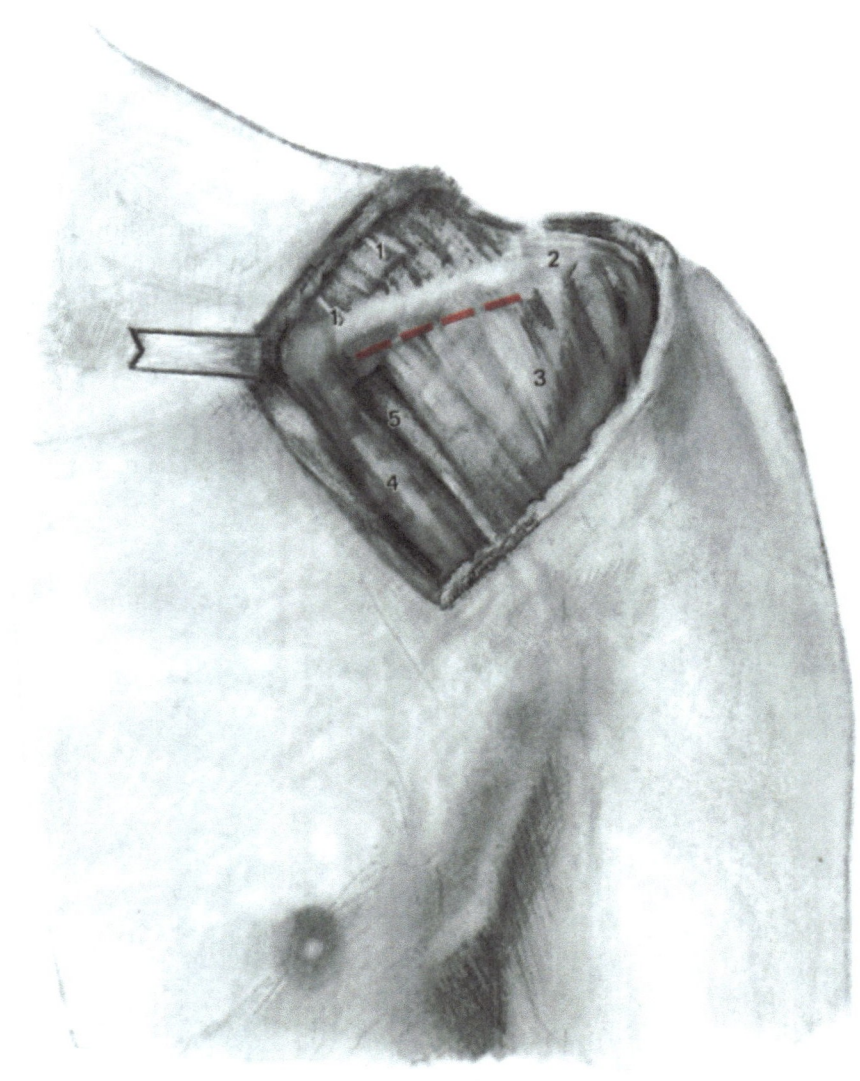

Deep Dissection (continued)   An intact acromio-clavicular (AC) joint is scarcely apparent on external inspection. With an AC dislocation, there is an obvious avulsion of the capsule from the clavicle or the acromion.

Internal Fixation   A 3.5 dynamic compression (DC) or reconstruction plate is applied to the clavicle, preferably to its antero-inferior border, as this provides a better hold for longer (3,5 mm) cortical screws and protrudes less beneath the skin.

Extension   Not required.

Closure   The deltoid and clavicular part of the pectoralis major are reattached.

Fig. 1c

1   Conoidal ligament
2   Trapezoidal ligament
3   Coraco-acromial ligament
4   Coracoid process
5   Pectoralis minor muscle

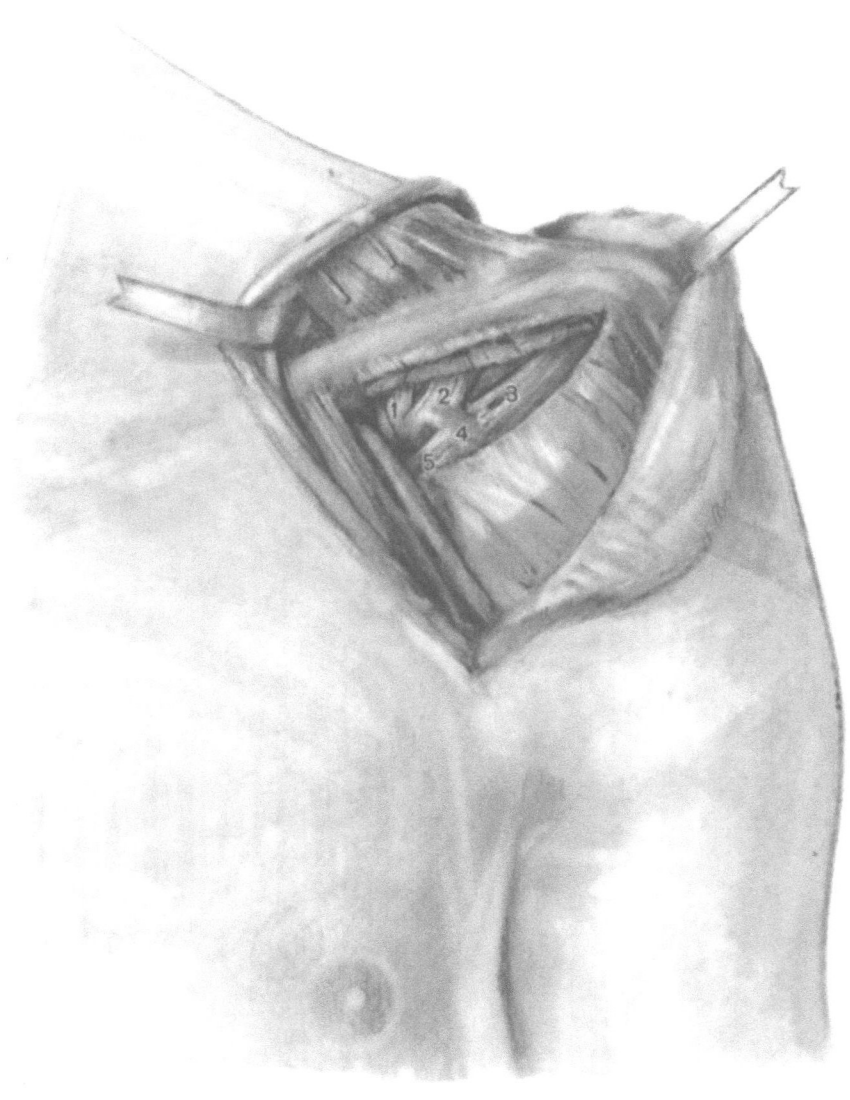

2 Scapula and Shoulder Joint: Posterior Approach

Indications   Displaced fractures of the shoulder joint, recurrent posterior dislocation of the shoulder.

Positioning/
Draping   Prone or lateral with the arm on a rest and draped so that it can be freely moved (see Chap. 5).

Skin Incision   Starts at the palpable posterior border of the *acromion*, passes along the *scapular spine*, and curves toward the *inferior angle of the scapula*.

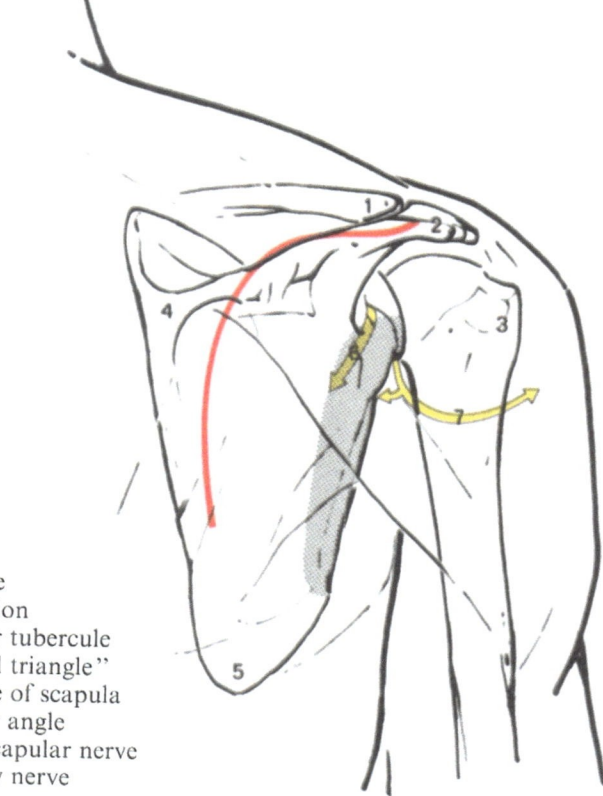

Fig. 2a

1 Clavicle
2 Acromion
3 Greater tubercule
4 "Spinal triangle"
    of spine of scapula
5 Inferior angle
6 Suprascapular nerve
7 Axillary nerve

Deep Dissection   The deltoid muscle is detached from the spine of the scapula, leaving a
narrow strip on the bone. Care is taken when reflecting the deltoid laterally,
as the axillary nerve and posterior circumflex humeral vessels emerging
from the quadrangular space course on its inferior surface.

Fig. 2b

1   Spine of scapula
2   Trapezius muscle
3   Deltoid muscle
4   Infraspinatus muscle
5   Teres minor muscle
6   Teres major muscle
7   Triceps muscle, long head
8   Superior lateral cutaneous nerve of arm (axillary nerve)

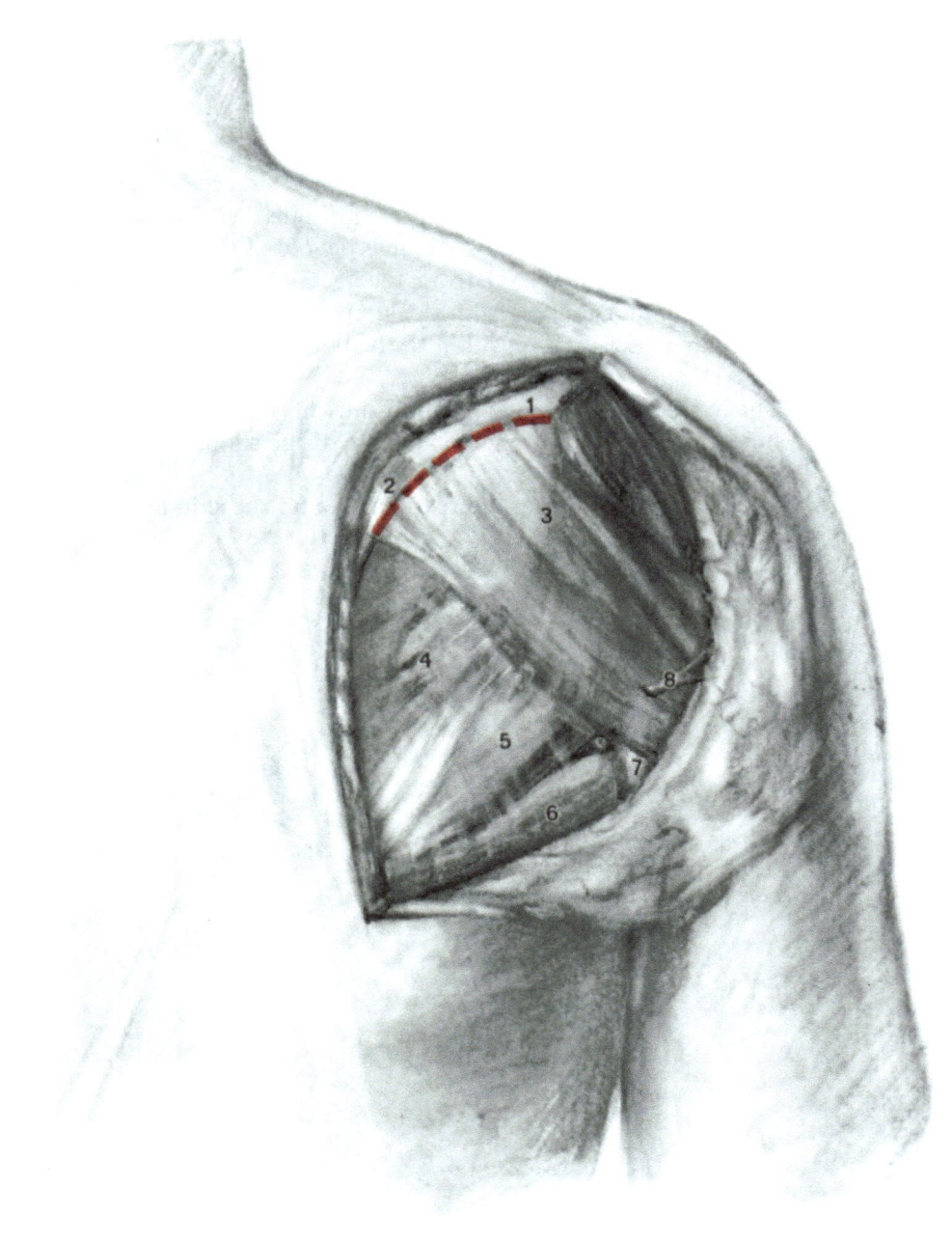

Deep Dissection
(continued)
The shoulder joint and inferior margin of the scapula are approached between the infraspinatus and teres minor muscles. Though somewhat difficult
to identify, this plane of dissection is devoid of nerves because the infraspinatus muscle is innervated from the medial side (suprascapular nerve) and
the teres minor muscle from the lateral side (axillary nerve). The circumflex
scapular vessels in apposition to the bone may have to be ligated.

Extension
The infraspinatus and teres minor muscles may be released from the greater
tubercule to obtain broader exposure of the shoulder joint.

Closure
All detached structures are reattached, and the skin is sutured.

Fig. 2c

1   Articular capsule
2   Deltoid muscle (reflected)
3   Circumflex vessels of scapula
4   Triceps muscle, long head
5   Axillary nerve and posterior
     circumflex humeral vessels
6   Teres minor muscle

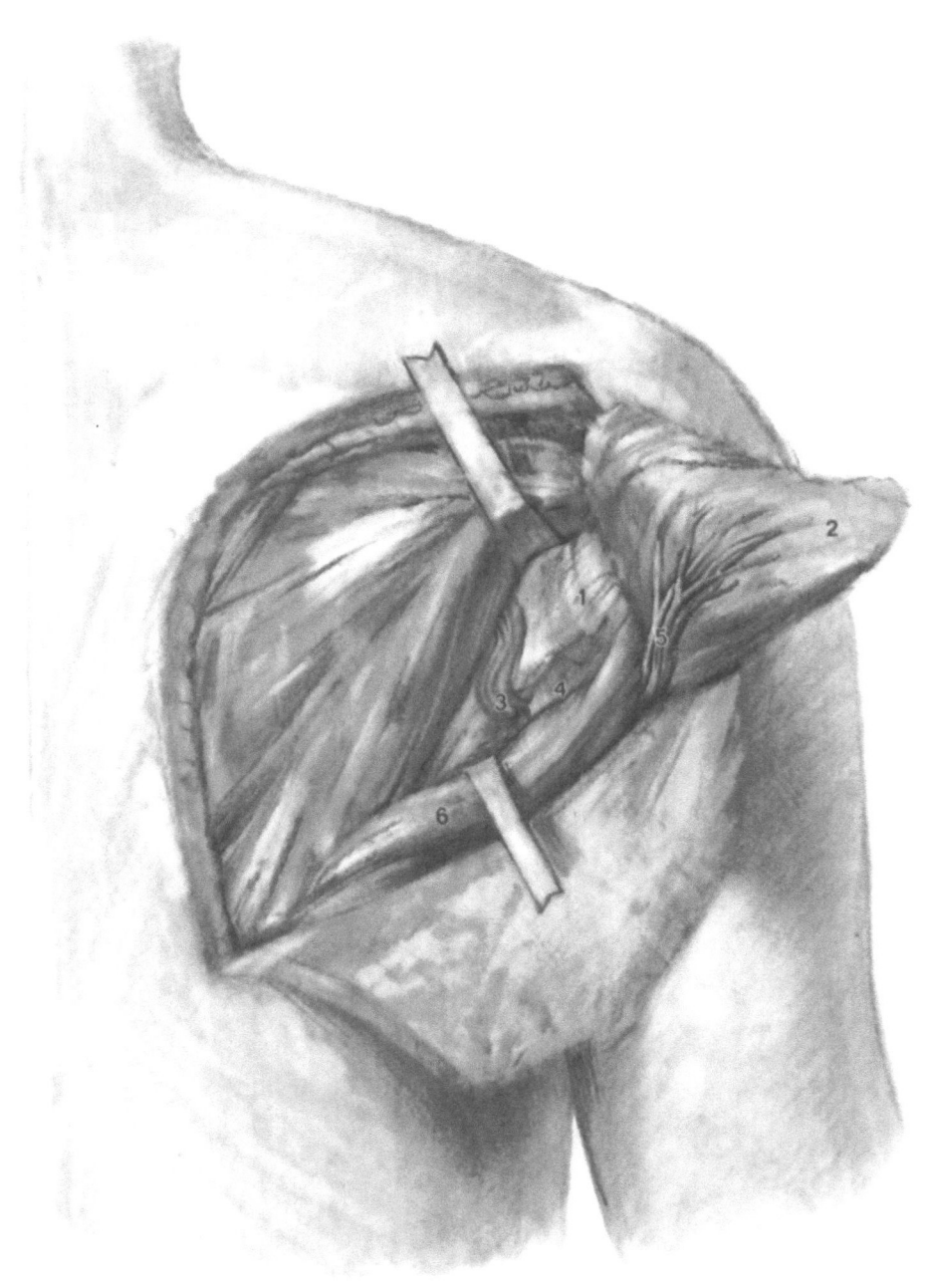

13

3 Proximal Humerus: Anterior Approach

Indications
: Severely displaced fractures or fracture-dislocations of the humeral head, operations on the shoulder joint.

Positioning/Draping
: Supine with ample padding beneath the shoulder (e.g., vacuum mattress); arm abducted 45°, head of the patient turned toward the opposite side.

Skin Incision
: From the *coracoid* process along the anterior border of the *deltoid muscle* to the *lateral bicipital sulcus*.

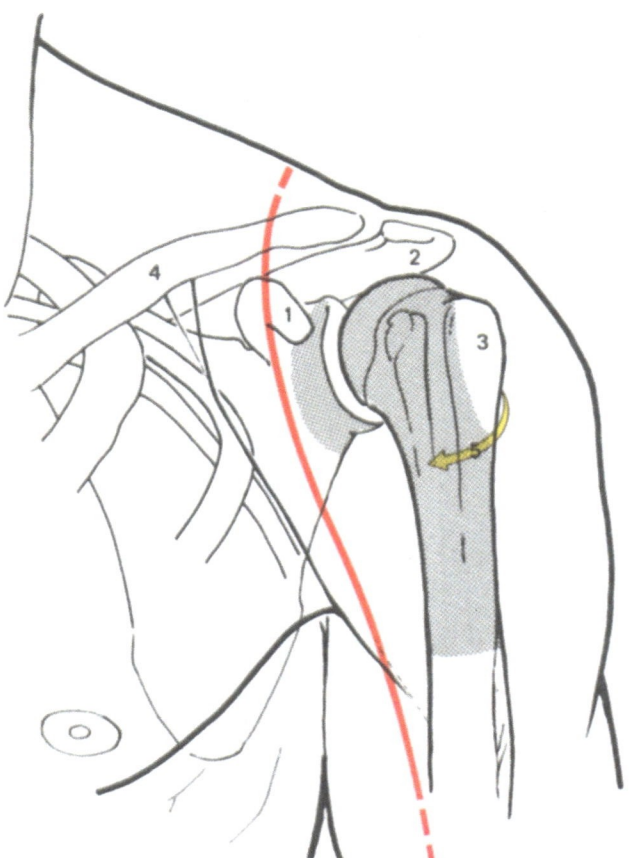

Fig. 3a

1 Coracoid process
2 Acromion
3 Greater tubercule
4 Clavicle
5 Axillary nerve

Deep Dissection   The cephalic vein is identified between the deltoid and pectoralis major. The two mucles are bluntly separated while the vein is retracted medially. This gives direct access to the proximal end of the humerus.

Extension   *Distally*
Along the cephalic vein into the lateral bicipital sulcus (see Chap. 4).

To the rim of the glenoid
The coraco-brachialis muscle and short head of the biceps are retracted medially. With the arm in slight external rotation, the subscapularis muscle is divided and retracted medially (with stay sutures), and the joint is opened. The tip of the coracoid process may have to be removed to improve access.

Fig. 3b

1   Deltoid muscle
2   Subscapularis muscle
3   Crest of greater tubercule
4   Biceps muscle, tendon of long head
5   Anterior circumflex humeral vessels
6   Pectoralis major muscle
7   Cephalic vein
8   Biceps muscle
9   Coracoid process with common flexor origin
10   Deltoid branch of the coraco-acromial artery

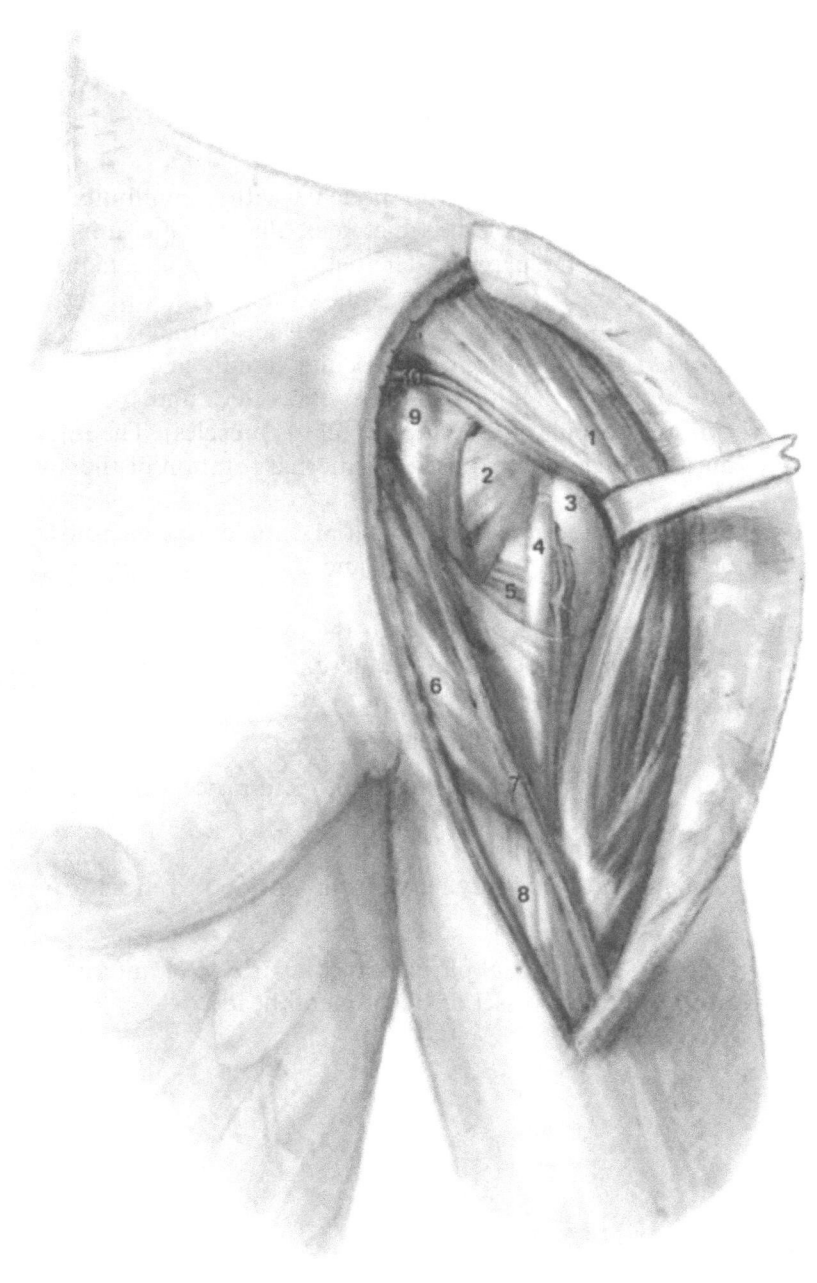

Extension (continued)

Laterally and proximally

In muscular individuals or patients with comminuted fractures, especially with involvement of the tubercules, the skin incision may be extended upward across the clavicle. The deltoid muscle is detached from the clavicle, leaving a narrow strip on the bone; osteotomy of the anterolateral tip of the acromion may also be necessary. The deltoid is carefully reflected, *taking care not to injure* the neurovascular bundle at its inferior surface, in order to expose the greater and lesser tubercules and the tendons of the rotator cuff (subscapularis and supraspinatus muscles). The infraspinatus and teres minor muscles are exposed by internal rotation of the humerus.

Internal Fixation

Plates and tension band material should not encroach upon the tendon of the long head of the biceps and should not be placed too far proximally.

Closure

– The deltoid is reattached to the clavicle and acromion (transosseous sutures). If the osteotomized acromion fragment is no wider than 1.5 cm, it is better to remove it than to attempt a reattachment.

– The subscapularis muscle is sutured.

– The osteotomized tip of the coracoid process is reattached with a screw.

Hazards

Lateral side

Damage to the axillary nerve from tension or retractor pressure.

Medial side

Injury to the brachial plexus, brachial artery, or anterior circumflex humeral artery (nutrient vessel of the humeral head).

Fig. 3c

1   Clavicle
2   Acromion
3   Lesser tubercule
4   Greater tubercule
5   Coracoid process

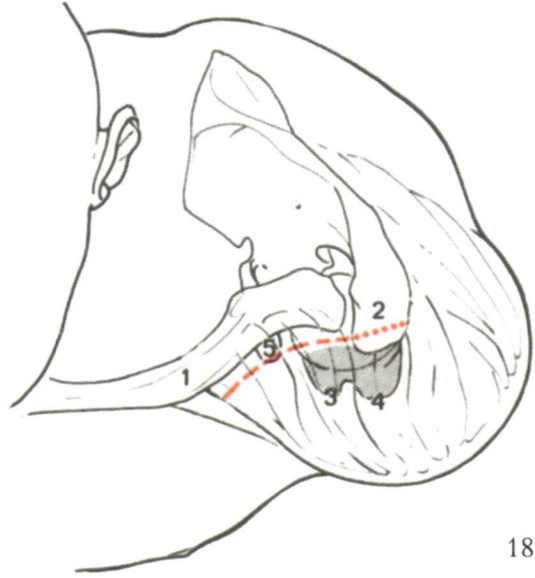

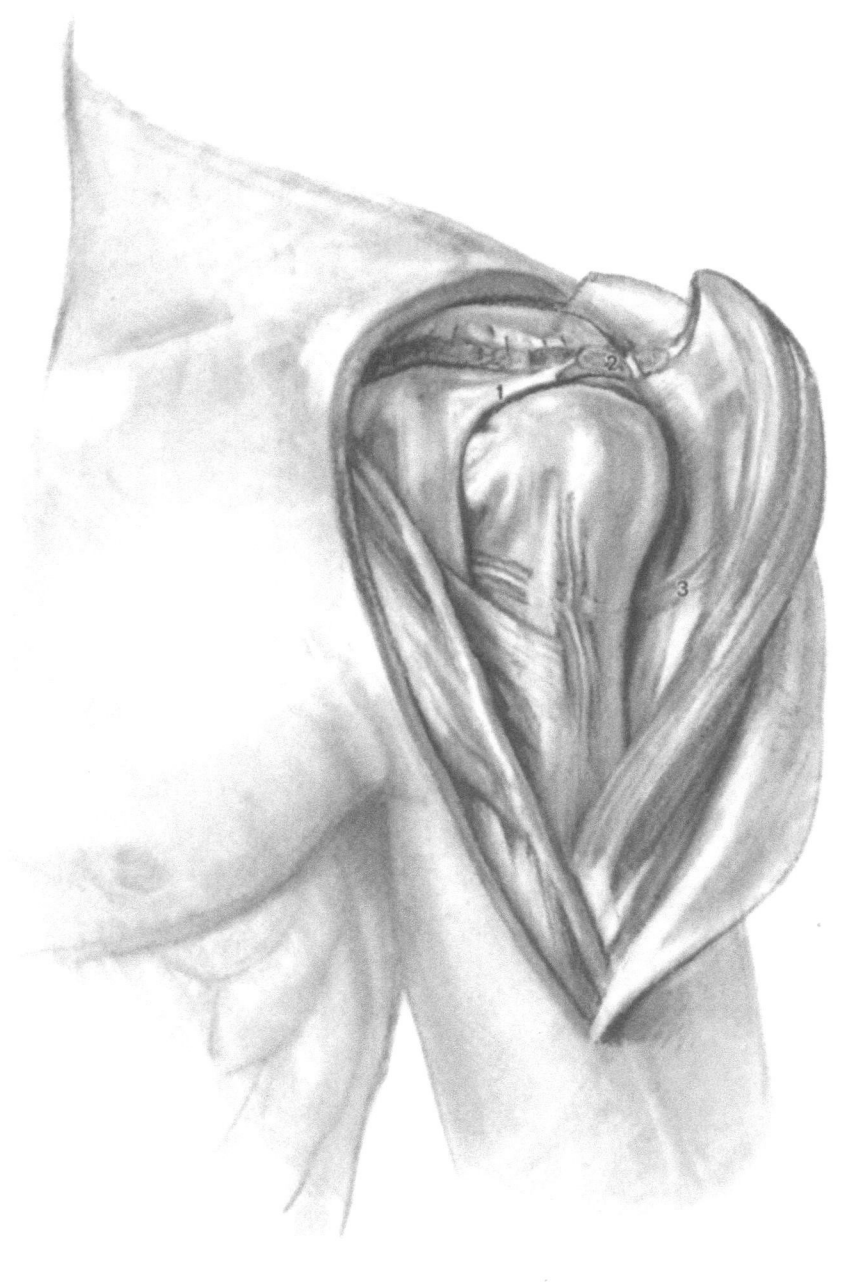

Fig. 3d

1   Coraco-acromial ligament
2   Acromion (after osteotomy)
3   Axillary nerve and posterior circumflex
     humeral vessels

4 Humeral Shaft: Antero-lateral Approach

Indications Fractures at the junction of the proximal and middle third of the humeral shaft.

Positioning/ Draping Supine with light padding beneath the shoulder; arm is supinated and abducted 60°–90°.

Skin Incision Starts 2 fingerwidths distal to the *coracoid process*, passes along the anterior border of the *deltoid muscle and the lateral bicipital sulcus*, and terminates 3 fingerwidths above the elbow.

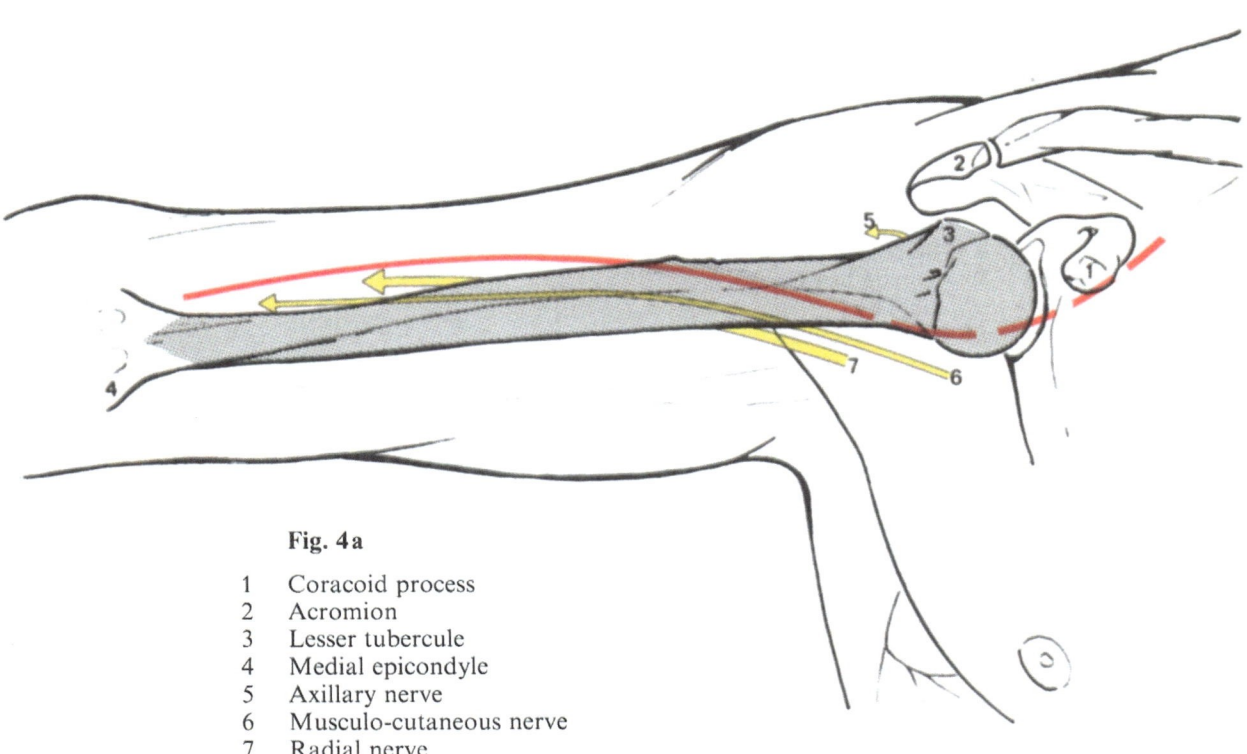

Fig. 4a

1 Coracoid process
2 Acromion
3 Lesser tubercule
4 Medial epicondyle
5 Axillary nerve
6 Musculo-cutaneous nerve
7 Radial nerve

4 Humeral Shaft: Antero-lateral Approach

Deep Dissection
The humerus is covered proximally by the deltoid muscle and distally by the brachialis muscle.

The bone is approached between the deltoid and pectoralis major muscles medial to the cephalic vein. The biceps muscle is retracted medially, and the brachialis is longitudinally split lateral to its midline; this is possible because its lateral portion is supplied by the radial nerve and its medial portion by the musculo-cutaneous nerve.

Fig. 4b

1	Biceps muscle	
2	Brachialis muscle	"Flexors"
3	Coraco-brachialis muscle	
4	Musculo-cutaneous nerve	
5	Brachial artery	
6	Median nerve	
7	Medial cutaneous nerve of the forearm	
8	Ulnar nerve	
9	Basilic vein	
10	Medial head of triceps muscle	
11	Long head of triceps muscle	"Extensors"
12	Lateral head of triceps muscle	
13	Radial nerve and profunda brachii vessels	
14	Cephalic vene	

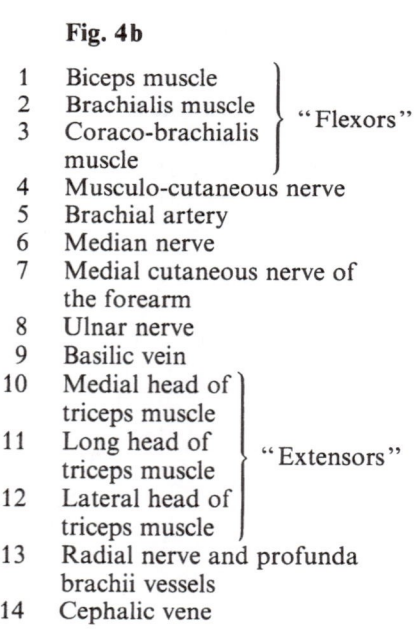
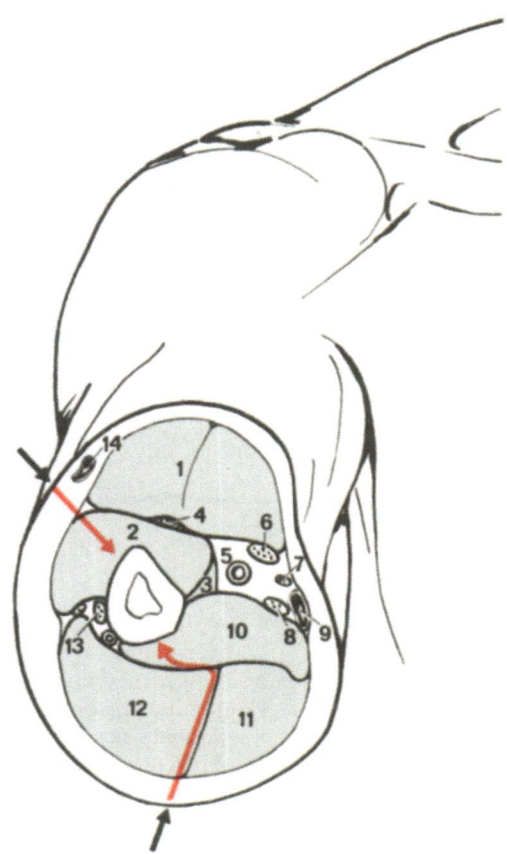

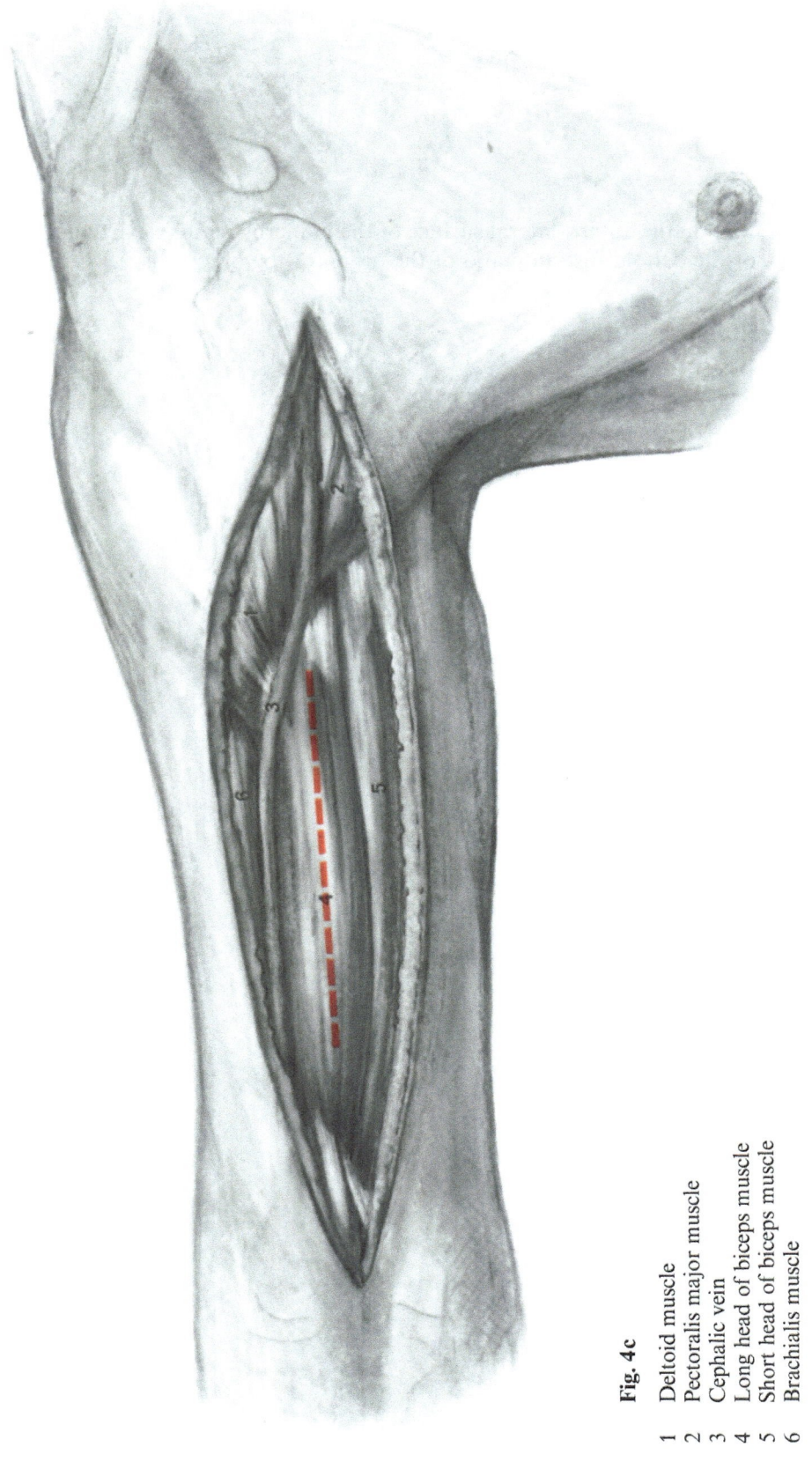

Fig. 4c

1 Deltoid muscle
2 Pectoralis major muscle
3 Cephalic vein
4 Long head of biceps muscle
5 Short head of biceps muscle
6 Brachialis muscle

Internal Fixation   Plating the antero-lateral surface of the humerus requires partial detachment of the deltoid insertion and of the origin of the brachialis.

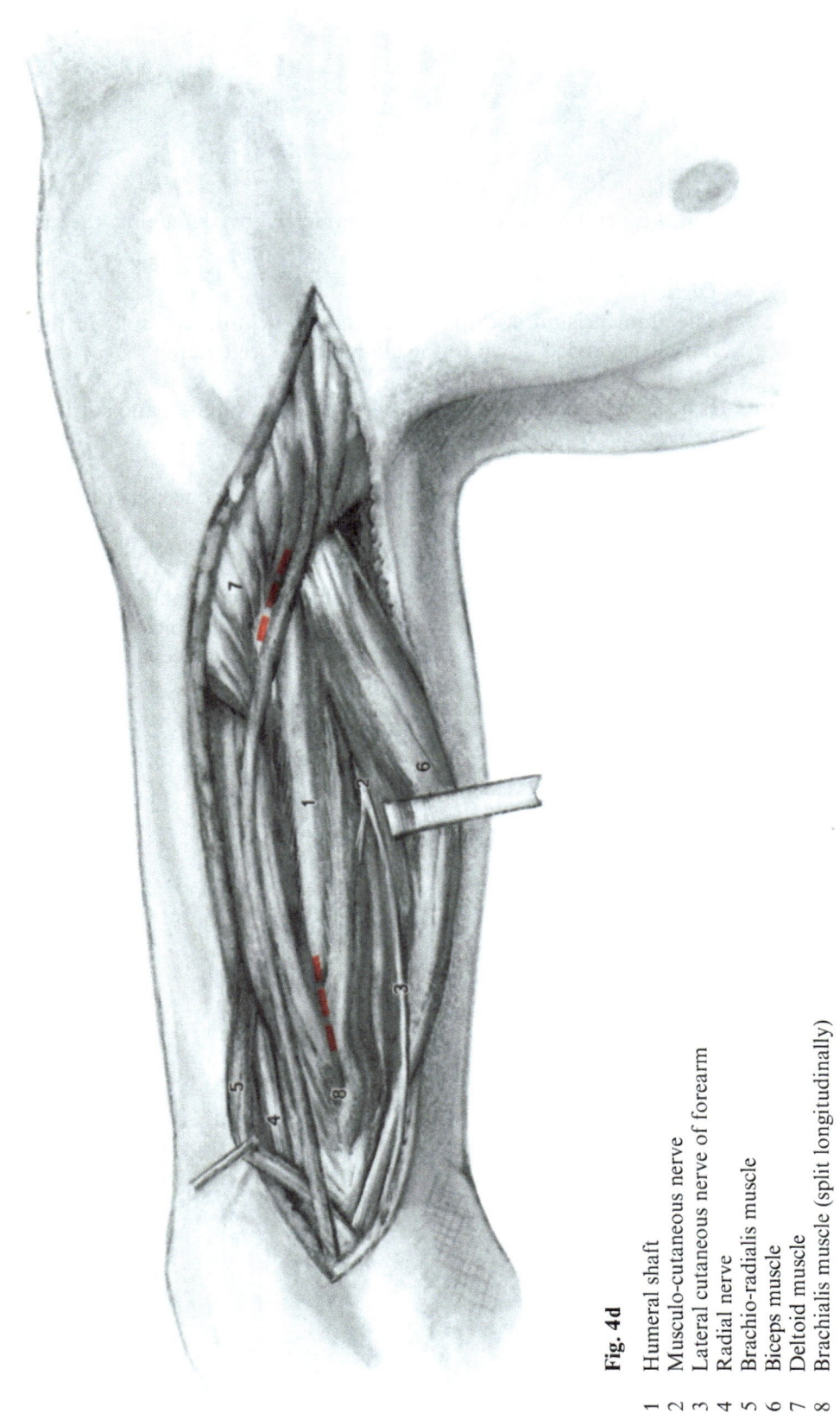

Fig. 4d

1 Humeral shaft
2 Musculo-cutaneous nerve
3 Lateral cutaneous nerve of forearm
4 Radial nerve
5 Brachio-radialis muscle
6 Biceps muscle
7 Deltoid muscle
8 Brachialis muscle (split longitudinally)

Extension   *Proximally*
The humeral head is exposed completely by detaching the deltoid from the clavicle (see Chap. 3).

Distally
The skin incision is extended to the elbow, and the split in the brachialis muscle is carried down to the elbow joint (see Chap. 8).

Closure   If necessary, the brachialis muscle is reapproximated and sutured.

Hazards   *Lateral side*
The radial nerve is in close proximity to the bone between the brachialis and brachio-radialis muscles.

Medial side
The lateral cutaneous nerve of the forearm (continuation of the musculo-cutaneous nerve) runs between the brachialis and biceps muscles. It enters the subcutaneous tissue at the lateral edge of the biceps tendon.

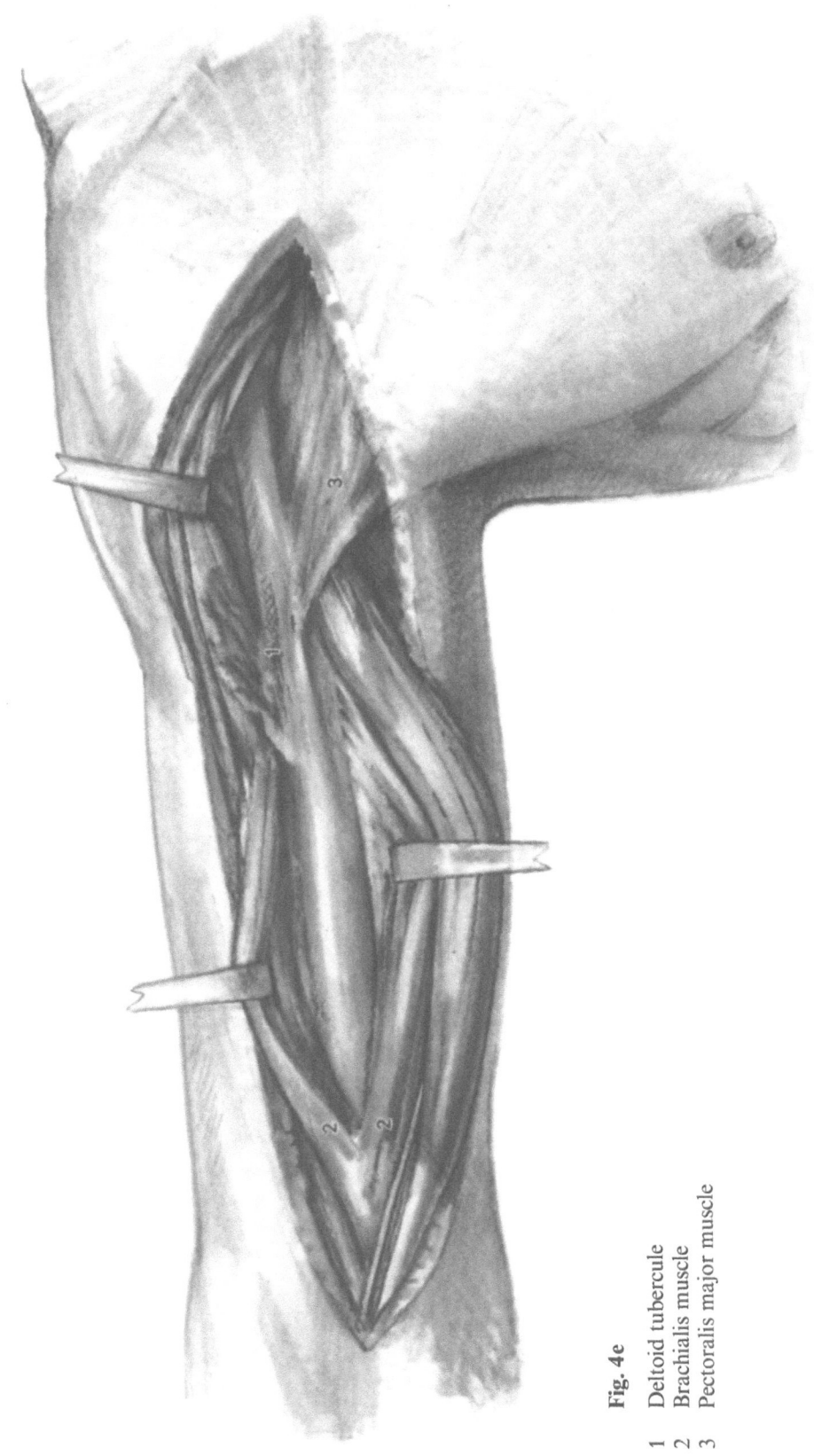

Fig. 4e

1   Deltoid tubercule
2   Brachialis muscle
3   Pectoralis major muscle

5 Humeral Shaft: Posterior Approach

Indications — Humeral shaft fractures of the middle and distal third.
Revision of the radial nerve.

Advantages — The radial nerve is visualized and can be mobilized or revised.

Positioning/Draping — Prone or lateral with the arm abducted 90° over an elbow rest. The elbow and shoulder are draped free.

Skin Incision — The posterior surface of the upper arm is incised along a line joining the *acromion* (posterior edge) with the *olecranon*, starting at the free border of the deltoid and ending 2 fingerwidths proximal to the tip of the olecranon.

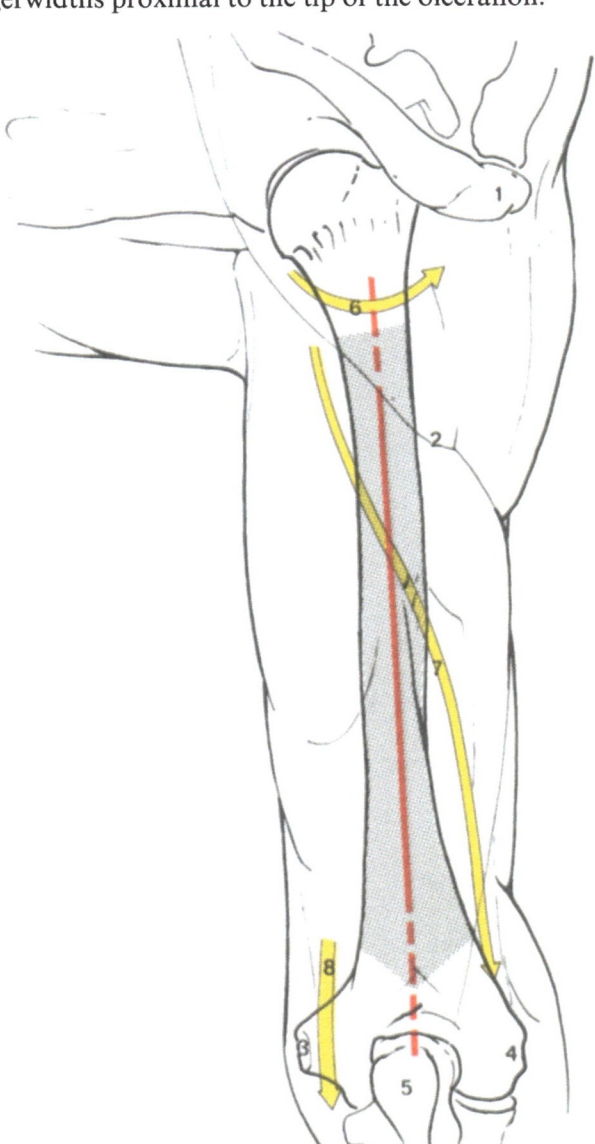

Fig. 5a

1 Acromion
2 Deltoid muscle (dorsal border)
3 Medial epicondyle of humerus
4 Lateral epicondyle of humerus
5 Olecranon
6 Axillary nerve
7 Radial nerve
8 Ulnar nerve

Deep Dissection   The posterior border of the deltoid is identified and the fascia is incised, *taking care not to injure* the upper lateral cutaneous nerve of the arm (branch of the axillary nerve). An index finger (arrow) is inserted into the interval between the two superficial heads of the triceps muscle (long and lateral heads); these are bluntly separated down to the aponeurosis. At the base of the interval are the radial nerve and the deep brachial vessels that descend obliquely and laterally *across the deep (medial)* head of the triceps.

Fig. 5b

1   Deltoid muscle
2   Superior lateral cutaneous nerve of arm (axillary nerve)
3   Lateral head of triceps muscle
4   Long head of triceps muscle
5   Tendon of triceps muscle
6   Olecranon
7   Interwal between lateral and long head of triceps muscle

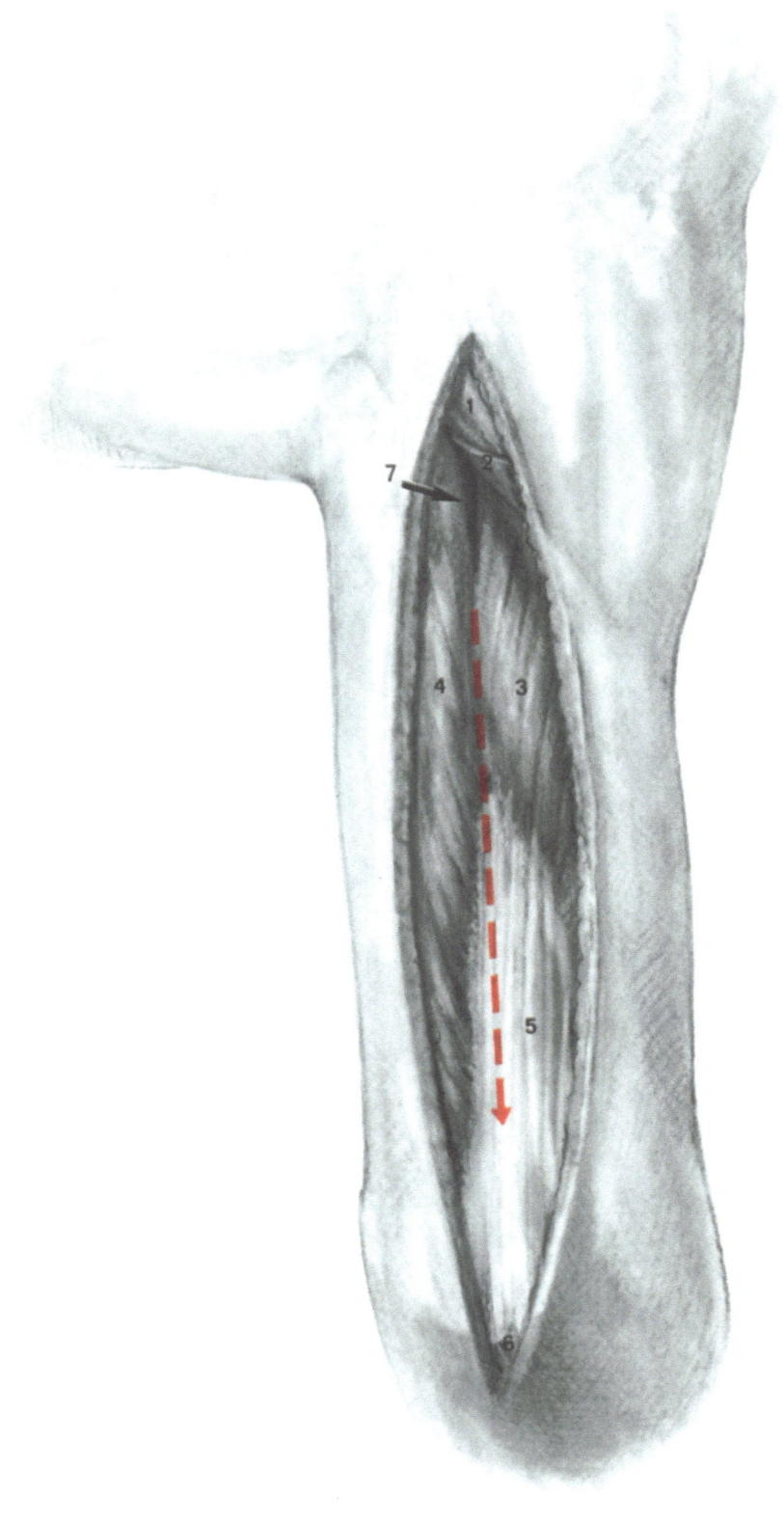

Deep Dissection
(continued)

By sharp dissection, the interval is extended down through the thick aponeurosis of the triceps to a point above the olecranon. After the radial nerve is snared with tape and retracted, the medial head is divided, and the humeral shaft is exposed proximally and distally. The lateral intermuscular septum may be notched or divided if additional room is needed for radial nerve retraction.

Fig. 5c

1   Radial nerve and vasa profunda brachii
2   Medial head of triceps muscle

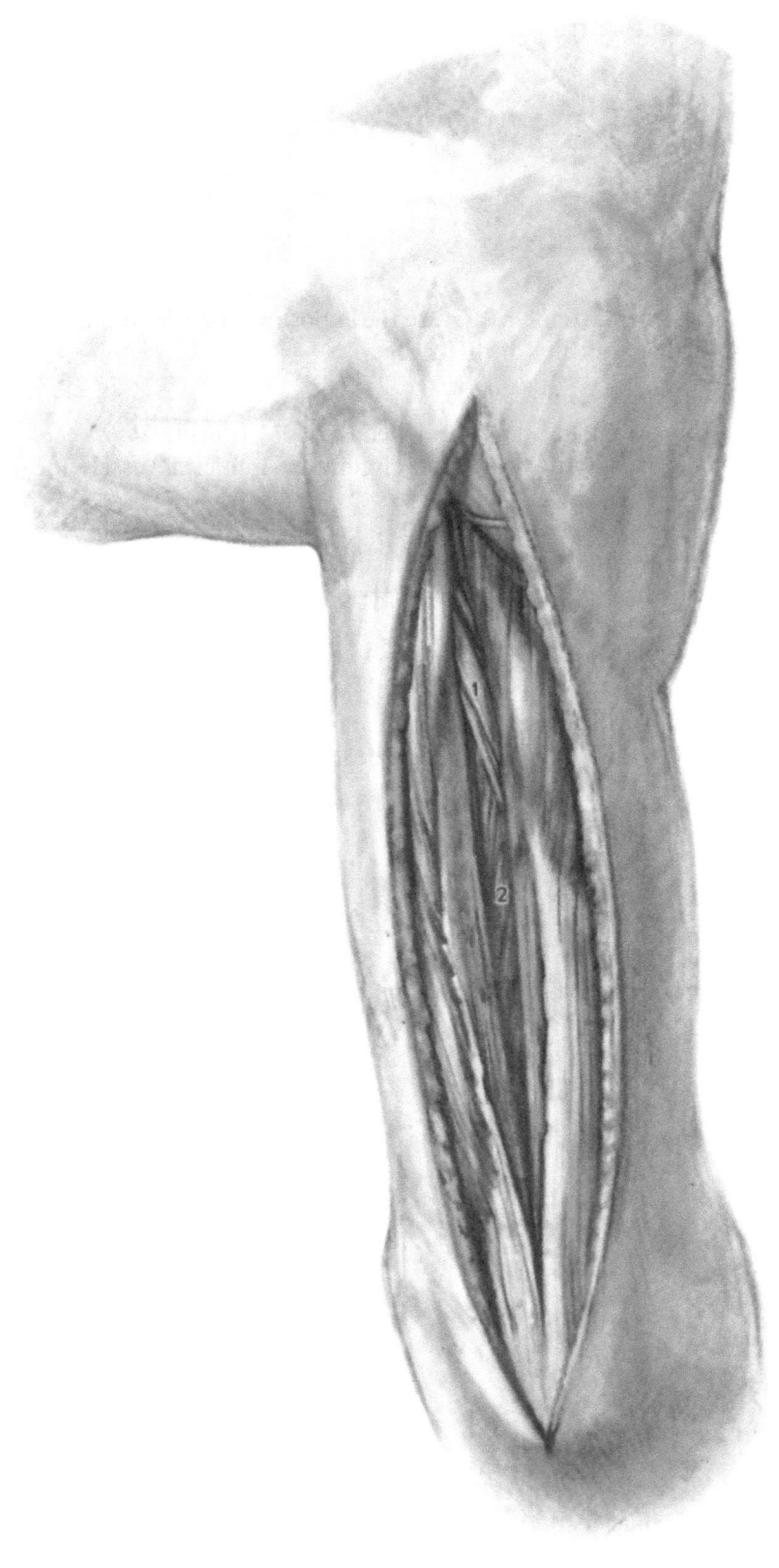

5   Humeral Shaft: Posterior Approach

Extension   *Distally*
For an intra-articular fracture the exposure can be extended to the olecranon by further splitting of the triceps.

Proximally
To the origin of the lateral head of the triceps. *Beware of injury* to the axillary nerve and posterior circumflex humeral vessels located proximal to it.

Internal Fixation   A tension-band plate is applied directly to the posterior surface of the humerus, where it is crossed by the radial nerve (*caution during plate removal!*).

Closure   The triceps aponeurosis is reapproximated.

Hazards   Damage to the ulnar nerve and main vascular bundle of the arm running anterior to the long and medial heads of the triceps muscle.

Fig. 5d

1   Profunda brachii artery
2   Radial nerve
3   Posterior cutaneous nerve of forearm
4   Muscular branches
5   Articular capsule
6   Lateral intermuscular septum

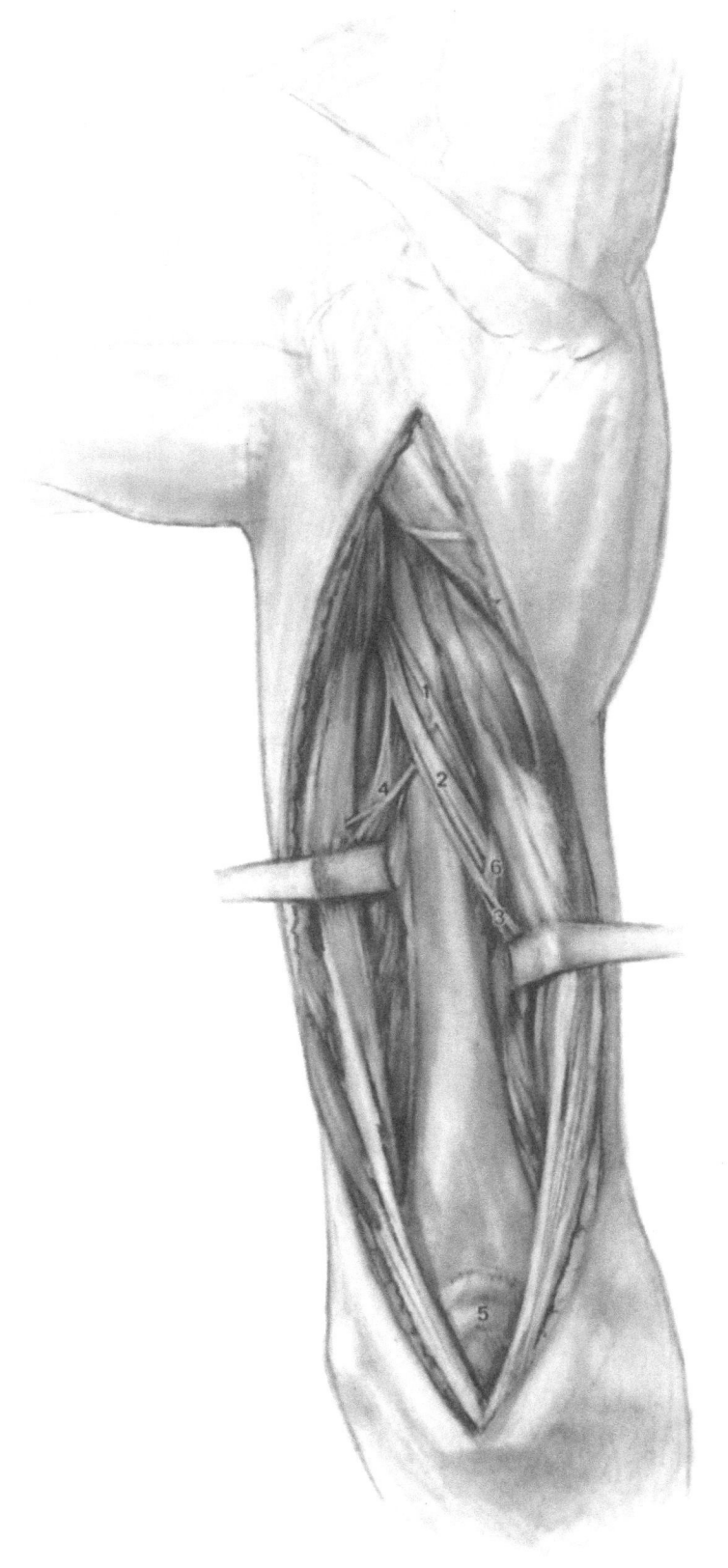

6 Distal Humerus and Humero-ulnar Joint

Indications   Distal intra-articular fractures of the humerus, fractures of the olecranon.

Positioning   *Intra-articular fractures of the humerus*
Prone or lateral with the arm abducted 90° on an elbow rest and the forearm hanging free; a sterile blood pressure cuff makes a suitable tourniquet.

Simple olecranon fractures
Supine with the arm laid across the chest.

Skin Incision   Starts over the mid-posterior surface of the distal *humerus* and curves *radially* past and 2 fingerwidths from the *olecranon* to the *posterior border of the ulna*.

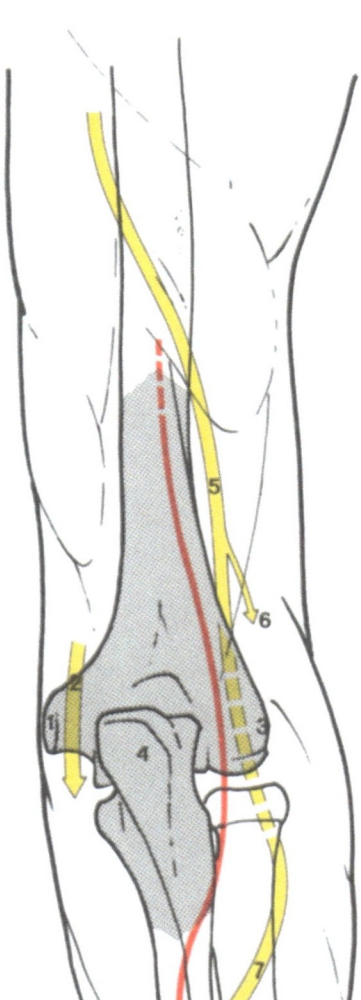

Fig. 6a

1 Medial epicondyle
2 Ulnar nerve
3 Lateral epicondyle
4 Olecranon
5 Radial nerve
6 Deep branch of radial nerve
7 Superficial branch of radial nerve

37

Deep Dissection   For fractures of the olecranon, simple medial reflection of the skin flap generally gives adequate exposure.

More complex intra-articular fractures require broad exposure of the distal humerus. This is done by osteotomizing the olecranon or dividing the triceps tendon in a V-shaped fashion.

Moreover, the ulnar nerve must be identified at the medial epicondyle of the humerus and snared with tape. In some instances a mobilisation of the nerve proximally and distally, as well as its anterior transposition, may be advisable.

The olecranon osteotomy should be made into the cartilage-free zone of the articular surface in such a way as to prevent rotation and to facilitate reattachment.

Fig. 6b

1   Triceps muscle
2   Ulnar nerve
3   Flexor carpi ulnaris muscle ("tendinous arch")
6   Olecranon
5   Anconeus muscle
6   Posterior crest of ulna

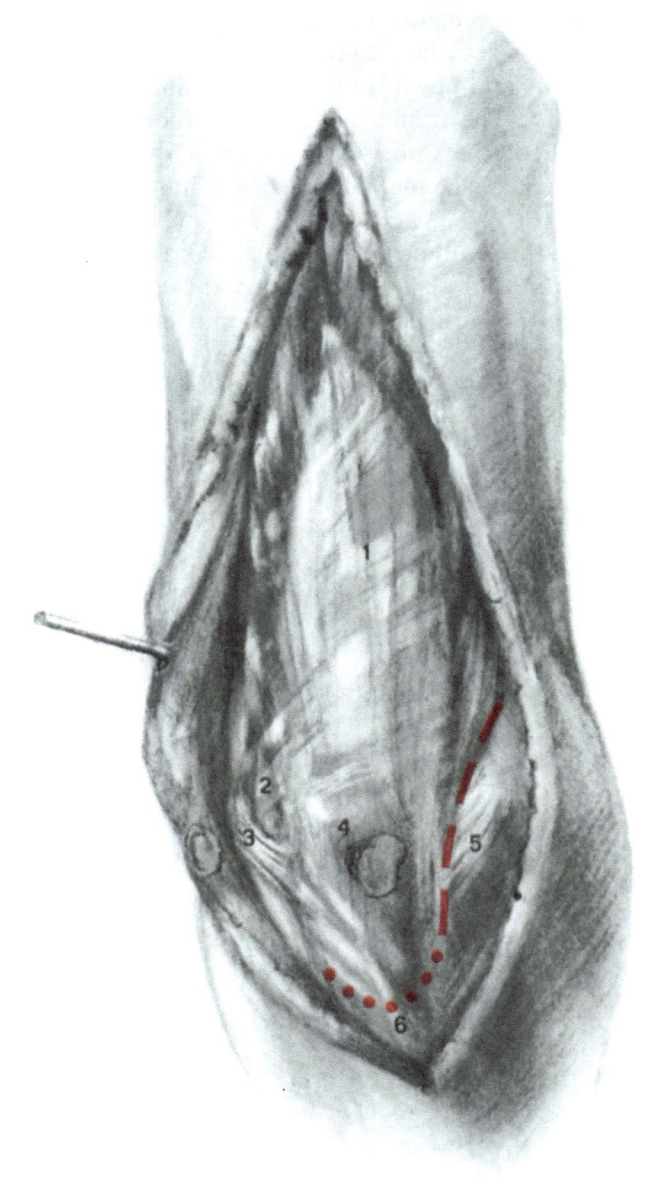

Deep Dissection (continued)	The rigid aponeurotic junction of the anconeus muscle is divided, and the triceps is detached from the lateral border of the humerus while pulling medially and upward on the osteotomized olecranon. *Caution:* The radial nerve pierces the lateral intermuscular septum in the middle of the upper arm. In most cases it is unnecessary to detach the triceps from the medial part of the humerus (*beware of* ulnar nerve injury).
Internal Fixation	A one-third tubular plate or 3.5 DC plate is applied to the posterior surface of the medial or lateral supracondylar ridges of the distal humerus, taking care not to obstruct the olecranon fossa.
Closure	The olecranon is reattached with a screw and tension band-wire. Anterior transposition of the ulnar nerve should be considered if internal fixation material has been inserted through the medial epicondyle.
Hazards	*Damage to radial nerve* in its path along the midportion of the humeral shaft.
	Damage to ulnar nerve at the medial humeral epicondyle.

Fig. 6c

1   Radial nerve
2   Triceps muscle
3   Lateral intermuscular septum
4   Posterior cutaneous nerve of forearm
5   Lateral epicondyle
6   Anconeus muscle
7   Olecranon (after osteotomy)
8   Ulnar nerve
9   Medial epicondyle

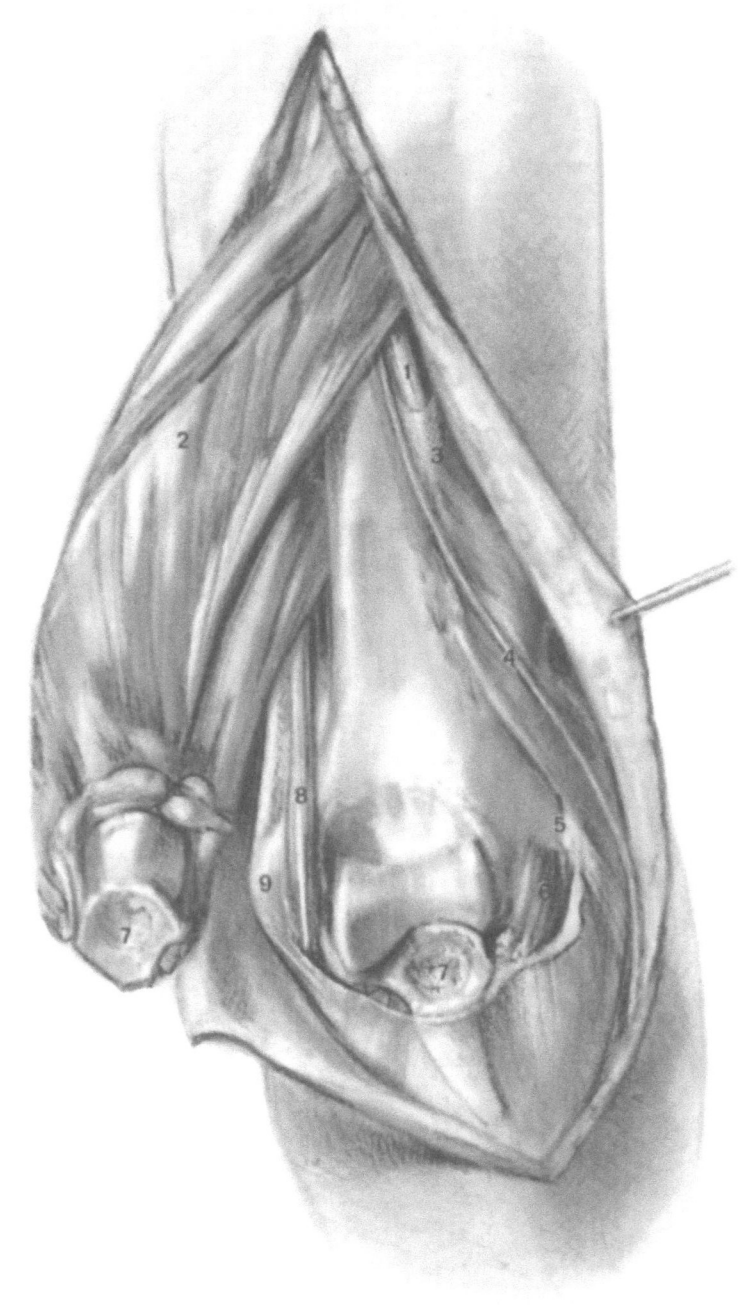

7 Humero-radial Joint and Radial Head

Indications
: Fractures of the head of the radius, MONTEGGIA's fractures, tears of the radial ligamentous apparatus.

Positioning/ Draping
: Supine with the arm on a side board, the elbow slightly flexed and the forearm pronated; tourniquet.

Skin Incision
: The dorsal incision starts 2–3 fingerwidths *proximal to the lateral epicondyle*, curves past the epicondyle toward the *styloid process of the ulna* and terminates about 3–4 cm distal to the humero-radial joint line.

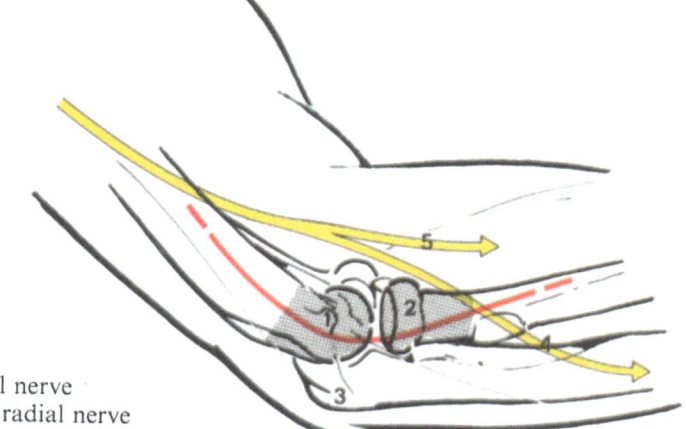

Fig. 7a

1 Lateral epicondyle
2 Head of radius
3 Olecranon
4 Deep branch of radial nerve
5 Superficial branch of radial nerve

7 Humero-radial Joint and Radial Head

Deep Dissection

The aponeurotic fascia over the anconeus muscle is split. The joint is approached through the interval between the anconeus and the common head of the extensor muscles. The lateral humeral epicondyle is either osteotomized and the common head of the dorsal extensor group (extensor digitorum communis, extensor digiti minimi, extensor carpi ulnaris) is reflected anteriorly and medially, or the common head is released from its attachment to the epicondyle. The humero-radial articulation is exposed by opening the joint capsule longitudinally, i.e., by incising the radial annular ligament and adjacent supinator muscle (*beware of injury* to the posterior interosseous nerve).

Fig. 7b

1   Triceps muscle
2   Posterior cutaneous nerve of forearm
3   Lateral epicondyle
4   Anconeus muscle
5   Common extensor origin

Fig. 7c

1   Triceps muscle, tendon
2   Anconeus muscle
3   Lateral epicondyle
4   Common extensor origin
5   Annular ligament
6   Interosseus recurrent artery
7   Posterior cutaneous nerve of forearm
     (radial nerve)
8   Radial extensor muscles

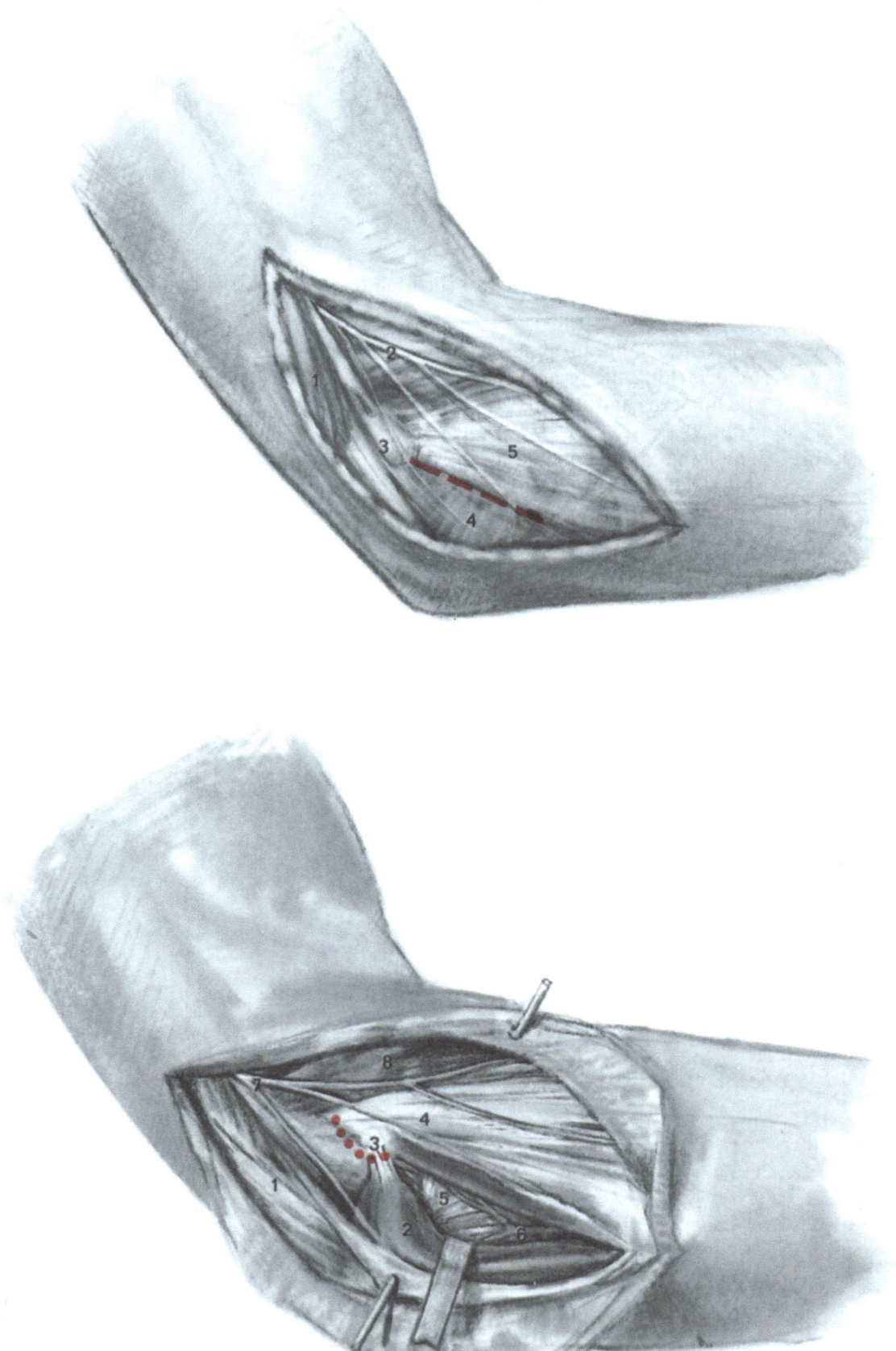

Extension   *Distally*
As needed along the border of the ulna (e.g., for a MONTEGGIA's fracture).

Medially
For reattaching an avulsed ulnar coronoid process, exposure is increased by detaching the brachioradialis and radial extensor muscles of wrist from the humerus and flexing the elbow joint.

Closure   The joint capsule, including the radial annular ligament, is closed with sutures. The osteotomized lateral epicondyle is reattached with a screw or with transosseous sutures for the common head of the dorsal extensor group.

Hazards   Damage to the posterior interosseous nerve within the supinator canal.

Fig. 7 d

1   Triceps muscle
2   Anconeus muscle
3   Head of radius
4   Lateral epicondyle (after osteotomy)
5   Annular ligament (cut)
6   Radial extensor muscles
7   Extensor digitorum muscle
8   Supinator muscle

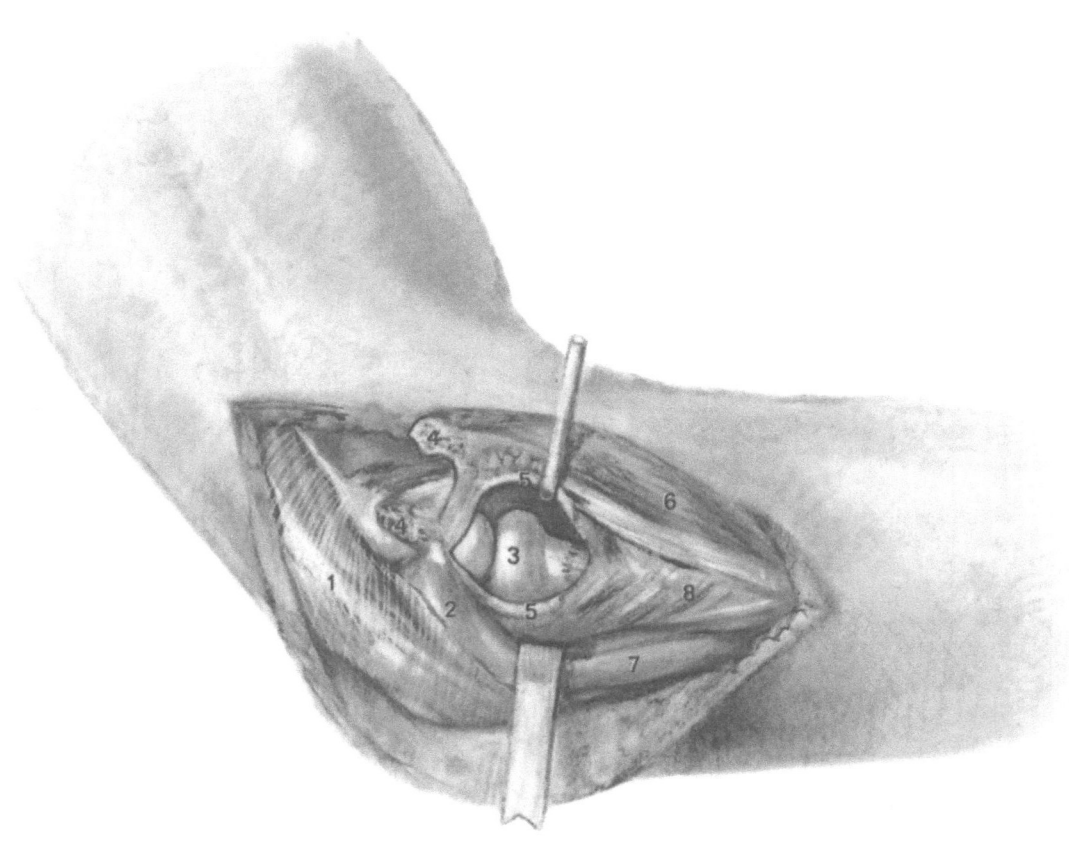

8 Elbow Joint and Proximal Radius: Volar Approach

Indication   Proximal radial fracture, biceps tendon avulsion from radial tuberosity, fracture of coronoid process, volar exposure of the humero-ulnar joint.

Positioning/ Draping   Supine with the arm abducted and supinated; tourniquet.

Skin Incision   An S-shaped incision 10–15 cm long is made from the *lateral bicipital sulcus* across the flexion creases to the *mid-forearm*.

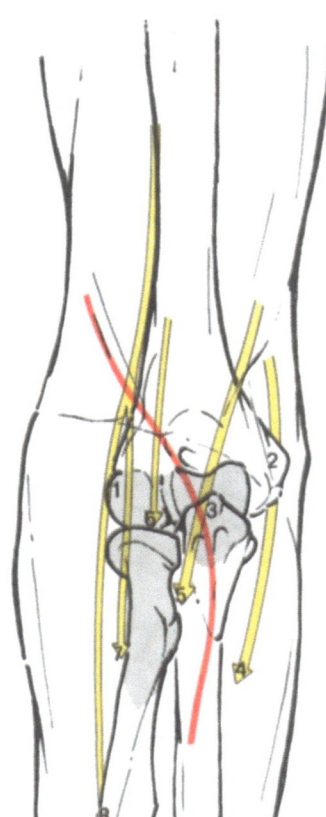

Fig. 8a

1   Capitulum of humerus
2   Medial epicondyle
3   Coronoid process
4   Ulnar nerve
5   Median nerve
6   Lateral cutaneous nerve of forearm
7   Deep branch of radial nerve
8   Superficial branch of radial nerve

Deep Dissection   To proximal end of radius and radial tuberosity.
The fascia is incised along the medial border of the brachio-radialis muscle.
Ligature of the superficial veins. (*Beware of* lateral cutaneous nerve of forearm.)

Fig. 8b

1   Median vein of forearm
2   Cephalic vein
3   Basilic vein
4   Lateral cutaneous nerve of forearm
5   Medial cutaneous nerve of forearm
6   Bicipital aponeurosis ("Lacertus fibrosus")

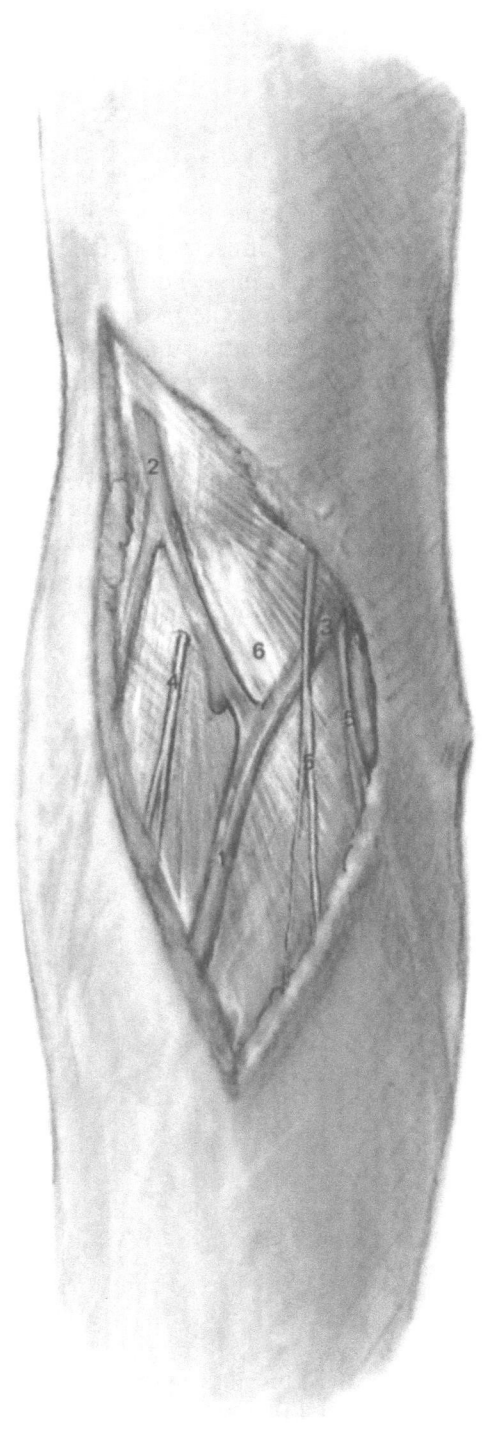

Deep Dissection
(continued)

The brachio-radialis is retracted laterally and the biceps tendon medially to expose the supinator muscle and the radial nerve, which divides into its superficial and deep branch. The fan-shaped leash of the recurrent radial artery, which curves around the biceps tendon, may be divided.
With the forearm in maximum supination, the supinator muscle is released from the proximal radius, *taking care not to injure* the deep branch of the radial nerve.

Extension

This exposure can be extended as far *distally* as the styloid process of the radius by releasing the supinator muscle (see above) and then retracting the radial extensor group (brachio-radialis, extensor carpi radialis longus et brevis) while the forearm is rotated to maximum *pronation*. The entire radial shaft can be exposed in this manner.

Fig. 8c

1 Brachio-radialis muscle
2 Supinator muscle
3 Biceps muscle
4 Pronator teres muscle
5 Tendon of biceps muscle
6 Bicipital aponeurosis (lacertus fibrosus)
7 Lateral cutaneous nerve of forearm
8 Deep branch of radial nerve
9 Superficial branch of radial nerve
10 Median nerve
11 Brachial artery
12 Radial artery
13 Radial recurrent artery
14 Joint capsule
15 Brachialis muscle

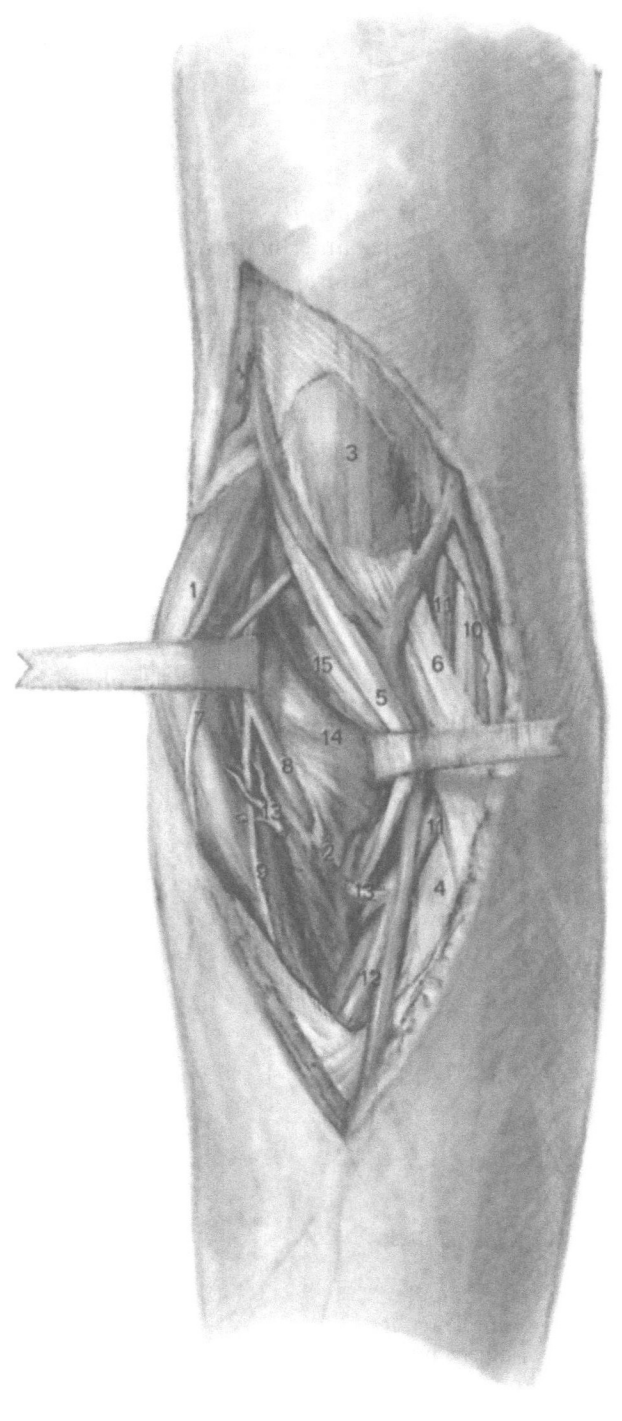

Deep Dissection (continued)

For surgery on the *coronoid process and humero-ulnar joint*
The superficial veins and bicipital aponeurosis ("lacertus fibrosus") are divided, sparing the lateral cutaneous nerve of the forearm. Deep to these structures the radial recurrent artery is ligated. The brachial artery, median nerve and pronator teres muscle are then retracted to the ulnar side, and the biceps tendon to the radial side. The fleshy brachialis muscle is split in line with its fibers to expose the coronoid process and capsule over the humero-ulnar joint.

Internal Fixation

A 3.5 DCP is applied to bone surface free of muscle.

Closure

The supinator is reattached if desired, the bicipital aponeurosis is reapproximated and sutured, and the skin is closed.

Hazards

The following structures are vulnerable to injury:

- the superficial branch of the radial nerve on the medial aspect of the radial extensors;

- the deep branch of the radial nerve, which is sandwiched in the "supinator canal";

- the median nerve and brachial artery deep to the bicipital aponeurosis.

Fig. 8d

1   Brachio-radialis muscle
2   Biceps muscle
3   Pronator teres muscle
4   Brachialis muscle (split)
5   Tendon of biceps muscle
6   Bicipital aponeurosis (lacertus fibrosus, cut)
7   Lateral cutaneous nerve of forearm
8   Superficial branch of radial nerve
9   Median nerve
10   Brachial artery
11   Radial artery
12   Ulnar artery
13   Radial recurrent artery
14   Coronoid process of ulna
15   Joint capsule

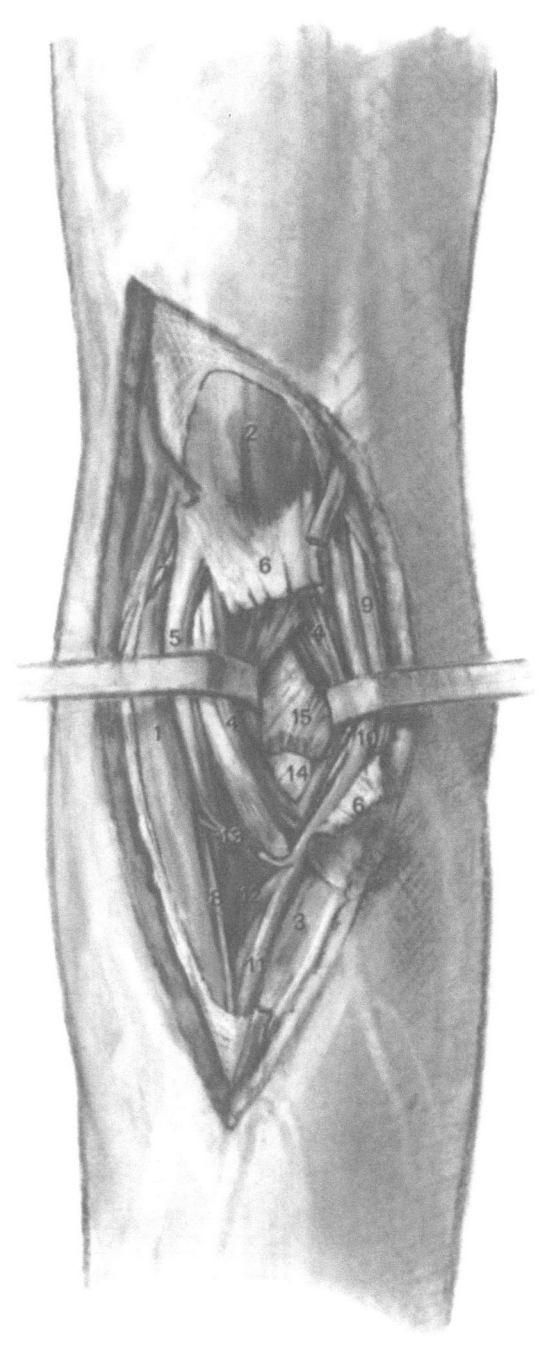

9 Radial Shaft: Dorso-lateral Approach

Indications	Radial shaft fractures, including fractures of the proximal and distal third.
Advantages	The posterior interosseous nerve may be safety identified, and a plate can be applied to the dorso-lateral, "tension-band" side of the radius.
Positioning/ Draping	Supine with the arm abducted and in neutral rotation; tourniquet.
Skin Incision	The *radial extensor group* is palpated (the "mobile wad" of Henry: extensor carpi radialis longus and brevis and brachio-radialis muscles). The incision is made on a line joining the *lateral epicondyle* of the humerus with the *radial styloid process*, along the dorsal border of the "mobile wad." The length corresponds to that of the fracture.

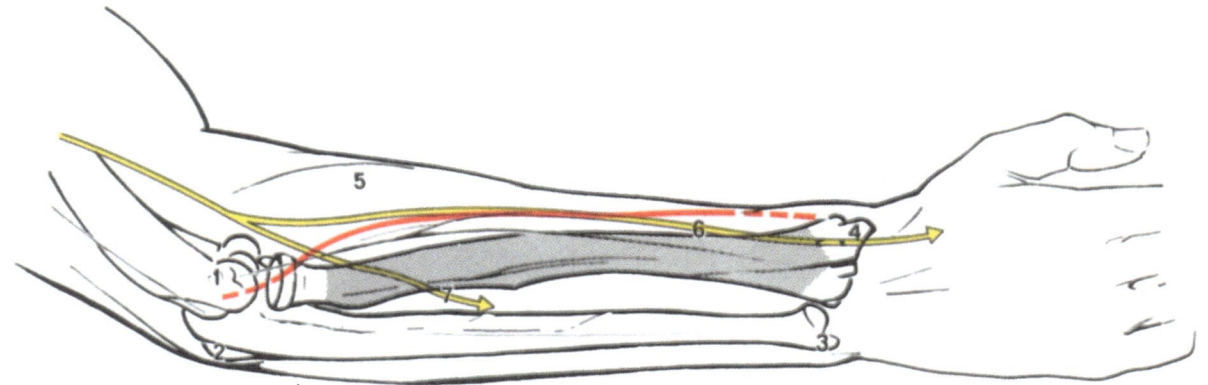

Fig. 9a

1 Lateral epicondyle of humerus
2 Olecranon
3 Styloid process of ulna
4 Styloid process of radius
5 "Mobile wad" (radial extensor muscles)
6 Superficial branch of radial nerve
7 Deep branch of radial nerve

Deep Dissection   *Distal third*

The bone is approached between the tendons of the radial extensor group and the muscle belly of the abductor pollicis longus (*beware of injury* to the superficial branch of radial nerve between the brachio-radialis and the extensors of the hand!).

Proximal third

The intermuscular septum, which is devoid of nerves, is split between the radial and dorsal extensor groups (between the "mobile wad" and the extensor digitorum communis) up to the lateral epicondyle of the humerus, exposing the supinator muscle over the proximal third of the radius.

Fig. 9b

1 Pronator teres muscle
2 Flexor carpi radialis muscle
3 Palmaris longus muscle
4 Flexor carpi ulnaris muscle
5 Flexor digitorum
   superficialis muscle   } "Flexors"
6 Flexor pollicis longus muscle
7 Flexor digitorum
   profundus muscle
8 Median nerve
9 Anterior interosseous nerve and vessels
10 Ulnar nerve and vessels
11 Brachio-radialis muscle
12 Extensor carpi radialis
    longus muscle   } "Superficial
13 Extensor carpi radialis   radial extensors"
    brevis muscle
14 Radial nerve and vessels
15 Extensor digitorum
    communis muscle
16 Extensor digiti   } "Superficial
    minimi muscle   dorsal extensors"
17 Extensor carpi
    ulnaris muscle
18 Abductor pollicis
    longus muscle
19 Extensor pollicis   } "Deep
    brevis muscle   extensors"
20 Extensor indicis
    longus muscle
21 Posterior interosseous
    nerve and vessels
22 Cephalic vein
23 Median vein of forearm
24 Basilic vein

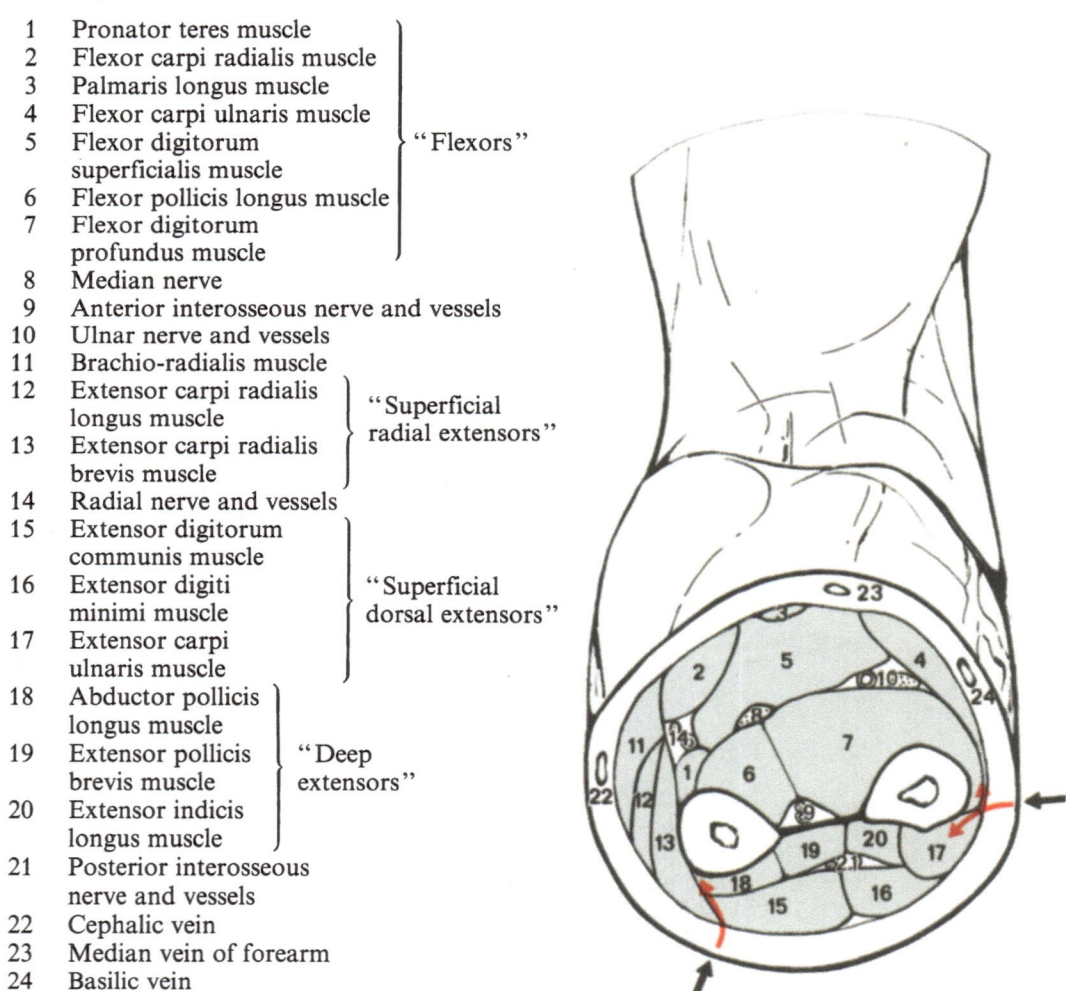

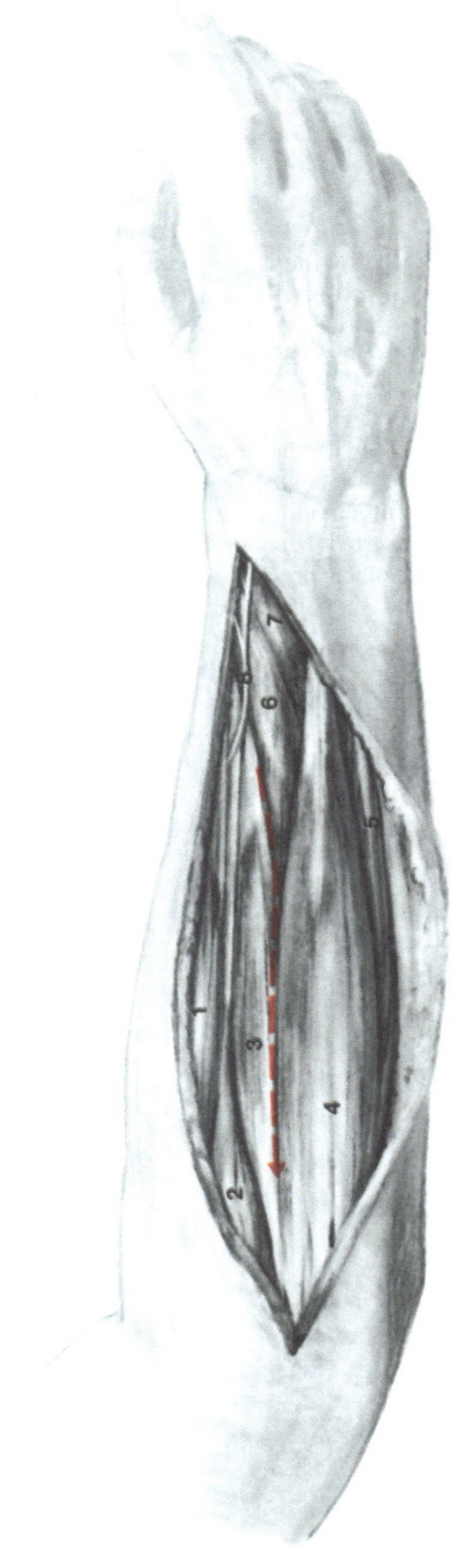

Fig. 9c

1 Brachio-radialis muscle
2 Extensor carpi radialis longus muscle
3 Extensor carpi radialis brevis muscle
4 Extensor digitorum muscle
5 Extensor carpi ulnaris muscle
6 Abductor pollicis longus muscle
7 Extensor pollicis brevis muscle
8 Superficial branch of radial nerve

Deep Dissection (continued)   The supinator is pierced by the deep branch of radial nerve (posterior inter-osseous nerve), which crosses the muscle at right angles to its fibers and is not accompanied by blood vessels. The nerve is palpable as a bulge about 3 fingerwidths distal to the radial head and can be demonstrated by separat-ing the muscle fibers.

Internal Fixation   To apply the 3.5 or 4.5 DCP, the supinator muscle together with the posteri-or interosseous nerve is detached from the radius, or the plate is inserted beneath the muscle and nerve in a distal-to-proximal direction. In the distal third, the plate is inserted beneath the tendons of the abductor pollicis longus and extensor pollicis brevis.

Extension   See "Deep dissection".

Closure   Generally a skin suture is sufficient.

Hazards   Injury to the superficial and deep branches of the radial nerve.

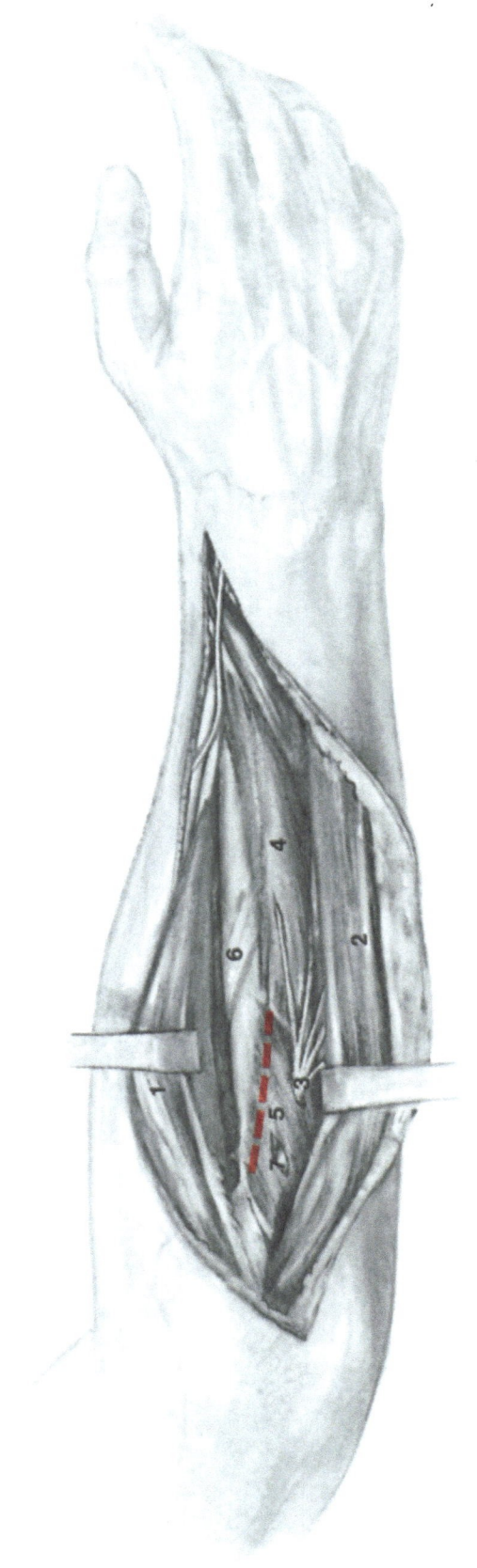

Fig. 9d

1 Extensor carpi radialis brevis muscle
2 Extensor digitorum muscle
3 Deep branch of radial nerve
4 Abductor pollicis longus muscle
5 Supinator muscle
6 Pronator teres muscle
7 Deep branch of the radial nerve within the supinator canal

10 Ulnar Shaft

Indications   Ulnar shaft fractures (for MONTEGGIA's fractures see Chap. 7).

Positioning/   Supine. With an isolated ulnar fracture, the arm is pronated and laid across
Draping   the chest. If both forearm bones are fractured, the arm is placed in abduction
and maximum pronation or the elbow is flexed.

Skin Incision   Parallel and slightly volar or dorsal to the palpable crest of the ulna.

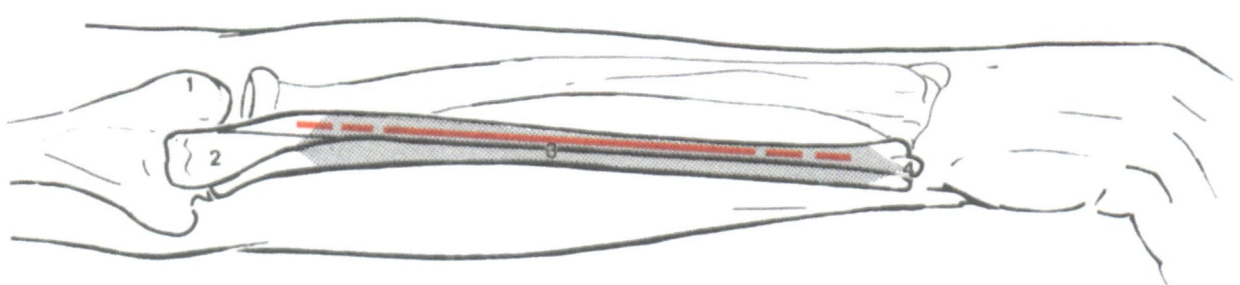

Fig. 10 a

1   Lateral epicondyle of humerus
2   Olecranon
3   Posterior crest of ulna
4   Styloid process of ulna

Deep Dissection   The flexors or extensors are detached from the dorsal crest of the ulna, depending on the course of the fracture line and the proposed site of the plate.

Internal Fixation   A 3.5 DCP is applied parallel to the ulnar border on the flexor or extensor side of the bone.

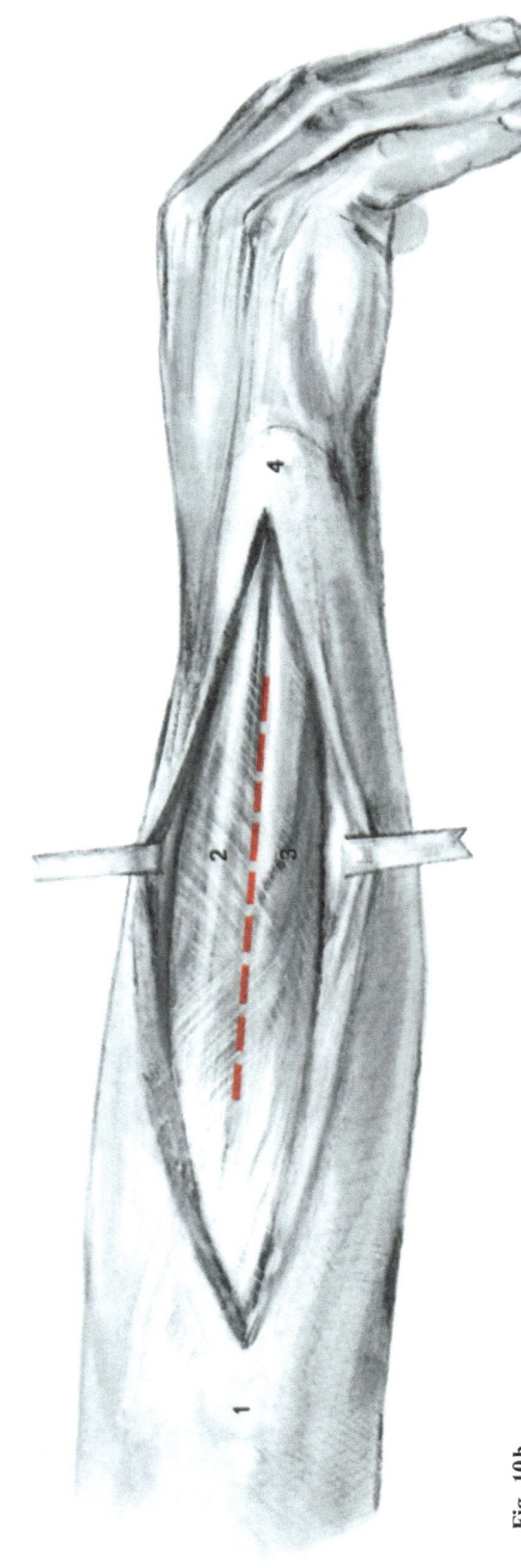

Fig. 10b

1 Olecranon
2 Extensor carpi ulnaris muscle
3 Flexor carpi ulnaris muscle
4 Head of ulna

10 Ulnar Shaft

Extension   *Proximally*
A curved extension is made past the olecranon and 1–2 fingerwidths from it on the medial or lateral side, depending on the primary incision (*beware of injury* to the ulnar nerve).

Distally
To the styloid process of the ulna.

Closure   The flexors or extensors are reattached and the skin is closed.

Hazards   Injury to the dorsal branch of the ulnar nerve between the bone and the flexor carpi ulnaris.

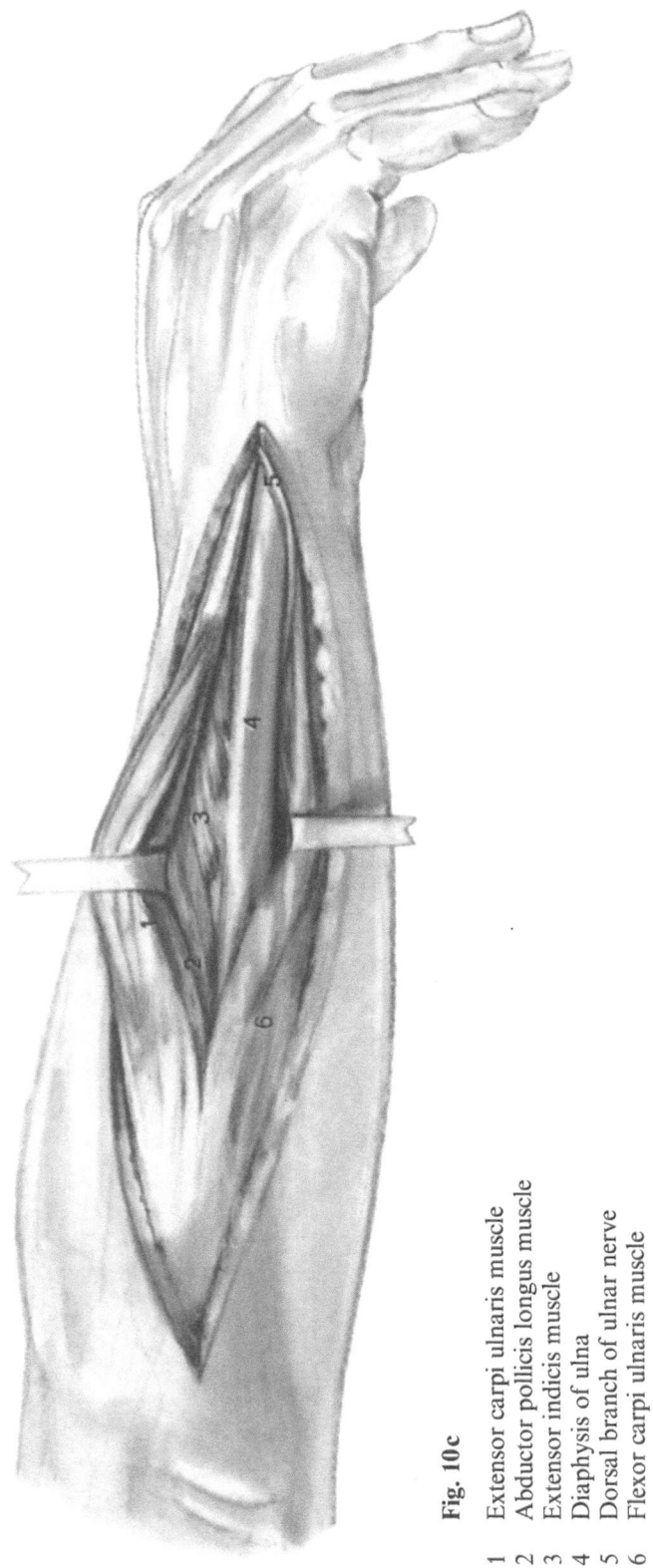

Fig. 10c

1 Extensor carpi ulnaris muscle
2 Abductor pollicis longus muscle
3 Extensor indicis muscle
4 Diaphysis of ulna
5 Dorsal branch of ulnar nerve
6 Flexor carpi ulnaris muscle

11 Distal Radius: Volar Approach

Indications | Intra-articular fractures of the distal radius (except those with dorsal displacement). Decompression of the median nerve.

Positioning/ Draping | The arm is abducted and supinated with the wrist extended over a roll; a tourniquet is applied.

Skin Incision | Starts along the *longitudinal crease* (linea vitalis), curves toward the ulnar side and back to the middle of the *forearm*. It extends 3–4 fingerwidths proximal and distal to the flexor crease of the wrist.

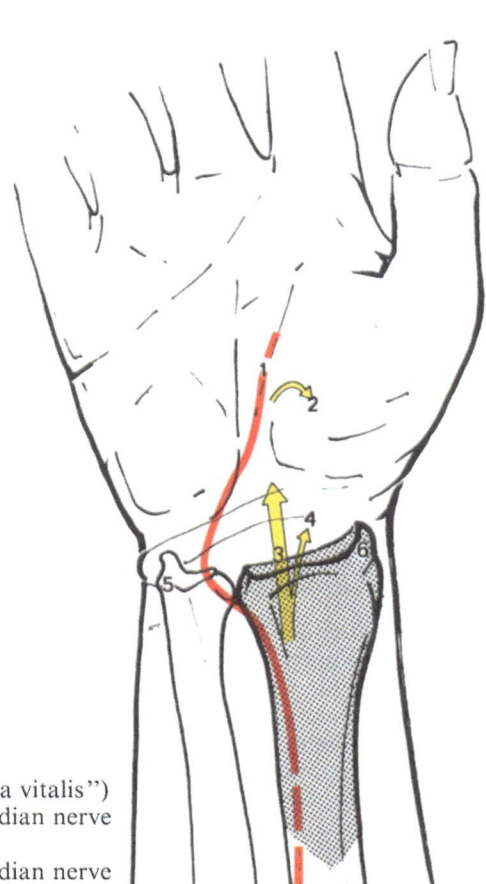

Fig. 11a

1　Thenar crease ("Linea vitalis")
2　Thenar branch of median nerve
3　Median nerve
4　Palmar branch of median nerve
5　Styloid process of ulna
6　Styloid process of radius

11 Distal Radius: Volar Approach

Deep Dissection   The flexor retinaculum is split medially to the tendon of the palmaris longus. This tendon, together with the median nerve and its sensory palmar branch, are retracted to the radial side.

Fig. 11 b

1 Palmar aponeurosis
2 Palmar branch of median nerve
3 Thenar branch of median nerve
4 Flexor retinaculum
5 Tendon of flexor carpi radialis
6 Tendon of palmaris longus
7 Tendon of flexor digitorum superficialis

Fig. 11 c

1 Palmaris brevis muscle
2 Flexor pollicis brevis muscle
3 Median nerve with "median artery"

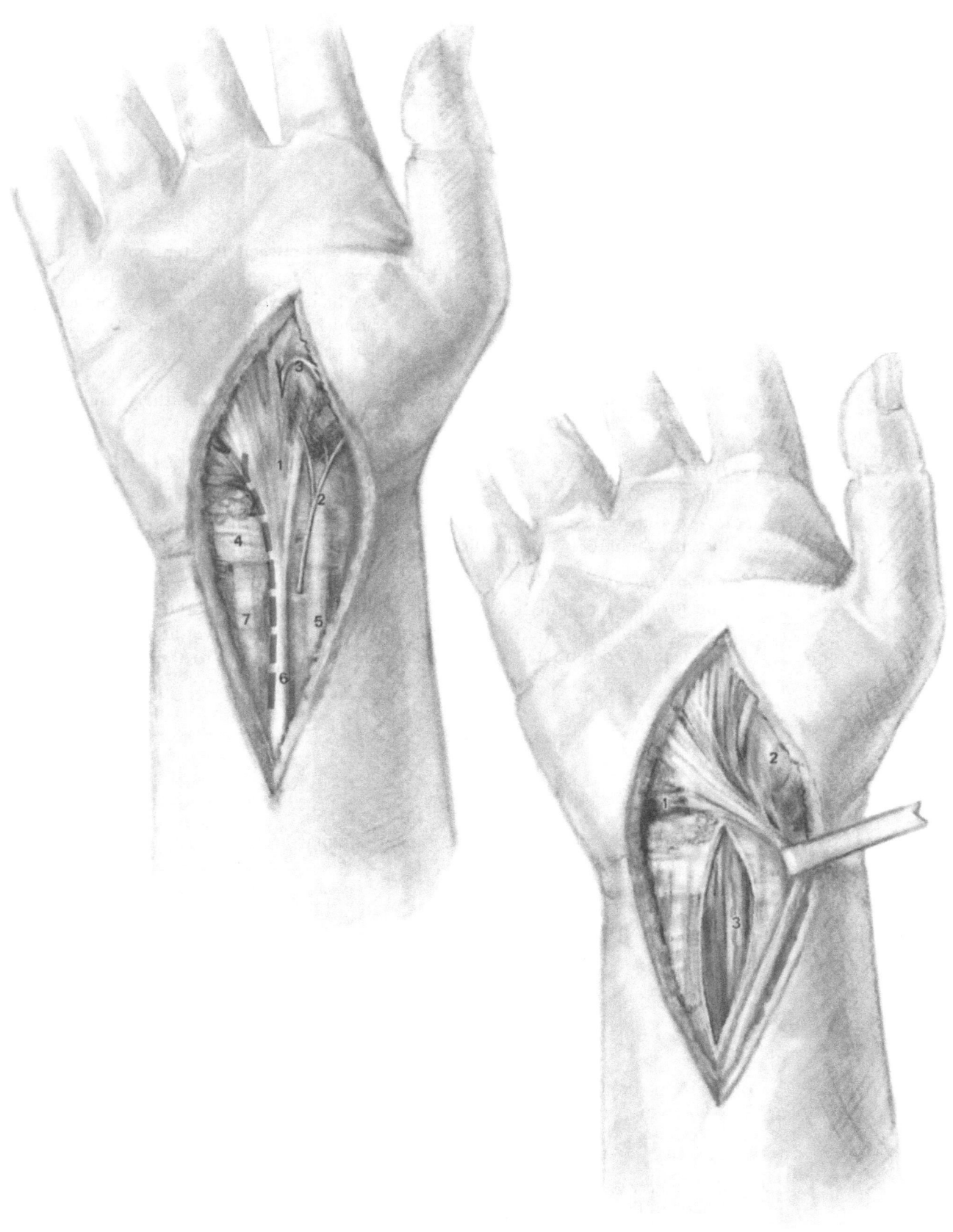

11 Distal Radius: Volar Approach

Deep Dissection (continued)   The distal end of the radius is approached between the tendons of the flexor carpi radialis and the digital flexors, but lateral to the flexor pollicis longus. The pronator quadratus muscle, which covers the bone, is notched in its distal portion or divided at its radial insertion and retracted to the ulnar side. The radiocarpal joint is opened transversely if need be.

Internal Fixation   A small-fragment T plate is applied to the volar surface of the radius to exert a buttressing effect.

Extension   *Distally*
The flexor retinaculum may be split further to decompress the median nerve (*spare* the palmar branch of the median nerve on the ligament and the thenar branch between the distal edge of the retinaculum and the thenar).

Toward the ulnar side
The group of digital flexor tendons may be retracted to the radial side, and the tendon of the flexor carpi ulnaris, the ulnar nerve, and the ulnar artery may be retracted to the ulnar side to gain access to the radioulnar joint.

Proximally
Along the midline of the forearm.

Closure   A skin suture is usually adequate.

Hazards   Injury to the median nerve or its sensory palmar branch.

Fig. 11 d

1 Median nerve
2 Tendon of flexor pollicis longus
3 Tendon of palmaris longus
4 Pronator quadratus muscle
5 Radiocarpal joint
6 Flexor digitorum superficialis and profundus

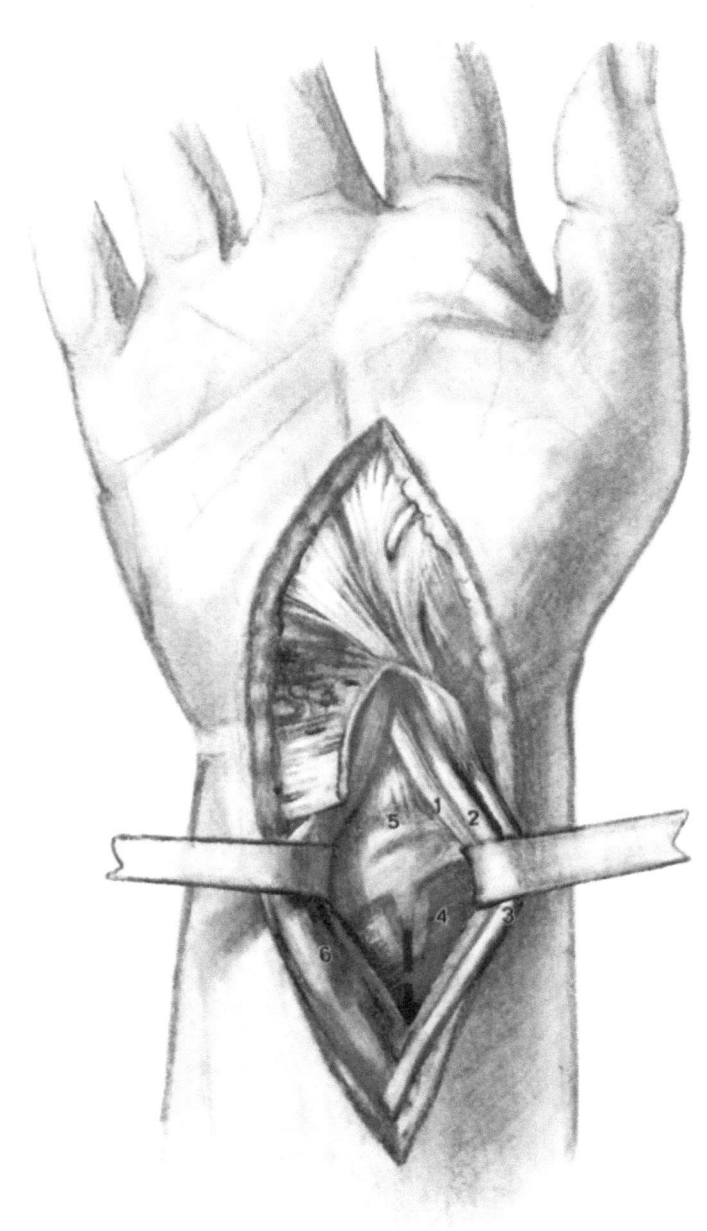

Part II

The Lower Extremities

12 Sacro-iliac Joint and Posterior Iliac Crest

Indications — Operations on the sacro-iliac joint or the gluteal aspect of the ilium. Removal of cancellous bone from the posterior iliac crest.

Positioning/ Draping — Lateral or prone such that the entire wing of the ilium is freely accessible; the anal cleft is well draped.

Skin Incision — Starts 1–2 fingerwidths distal and lateral to the *posterior superior iliac spine*, extends upward parallel to the *iliac crest*, and terminates 2–3 cm past the highest point of the crest. The superior clunial nerves must be divided.

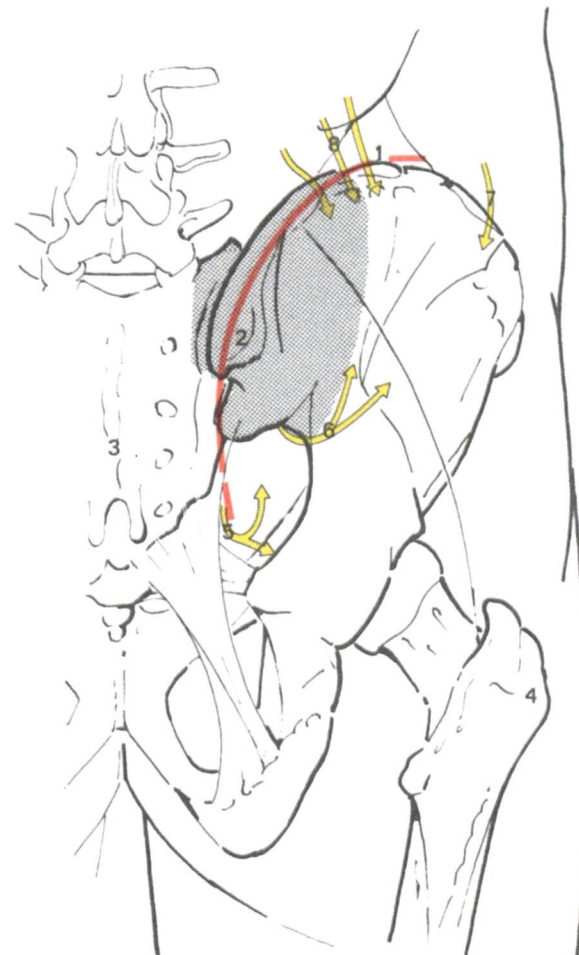

Fig. 12a

1 "Supracristal point" (highest point of iliac crest)
2 Posterior superior iliac spine
3 Medial sacral crest
4 Greater trochanter
5 Inferior gluteal nerve
6 Superior gluteal nerve
7 Lateral cutaneous branch of ilio-hypogastric nerve
8 Superior clunial nerves

Deep Dissection   The gluteus maximus muscle is divided near its attachment to the outer surface of the ilium and is carefully reflected downward. On its posterior surface, the superficial branches of the superior gluteal vessels and one branch of the inferior gluteal nerve emerge from the pelvis at the greater sciatic notch.

Fig. 12b

1   Lateral sacral crest
2   Lateral cutaneous branch of ilio-hypogastric nerve
3   Superior clunial nerves
4   Lumbar trigone (Petit's triangle)
5   Erector spinae muscles
6   Posterior superior iliac spine
7   Medial sacral crest
8   Gluteus maximus muscles
9   Ilio-tibial tract

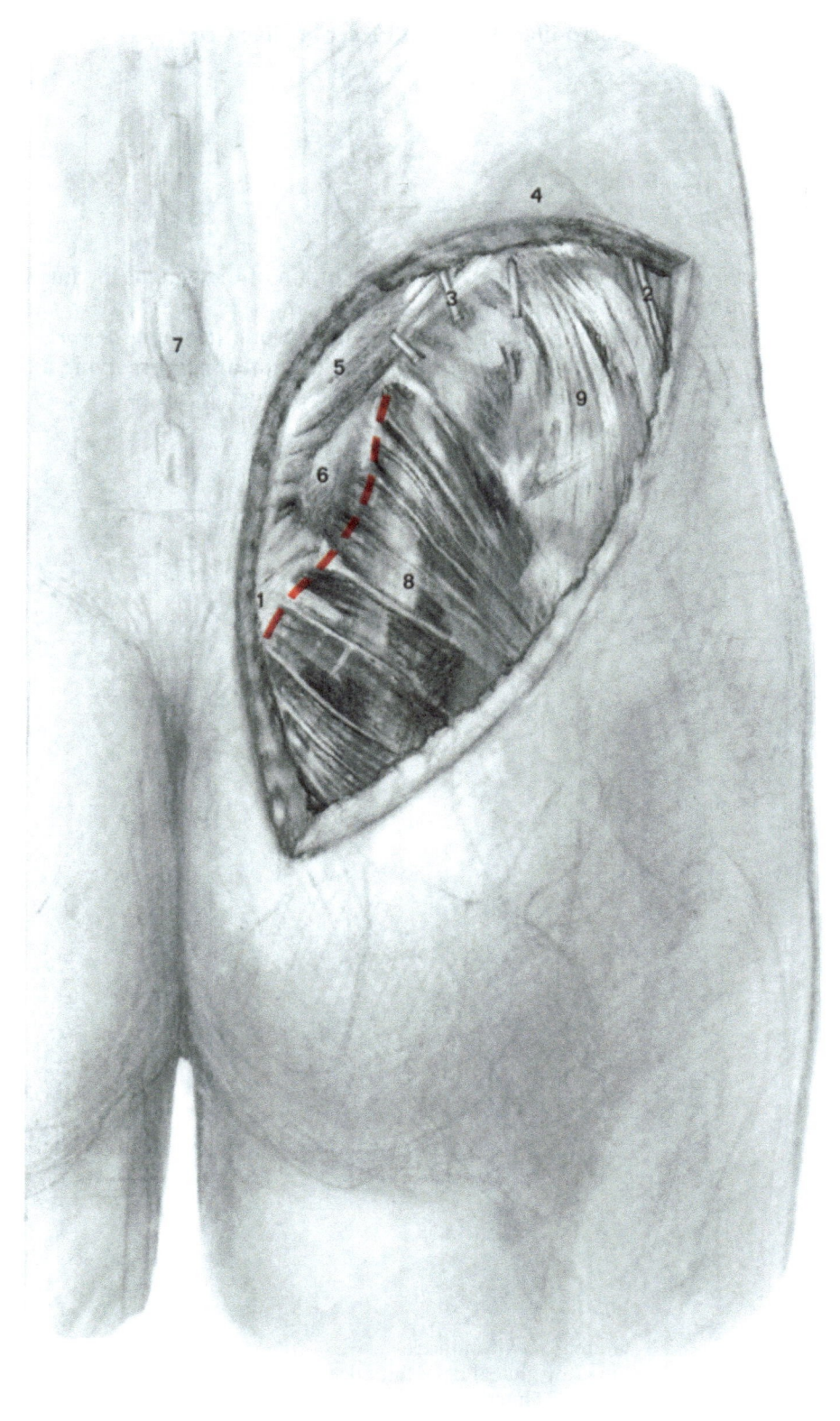

Deep Dissection
(continued)

If necessary, the posterior origin of the gluteus medius is reflected anteriorly.
The suprapiriform foramen may be enlarged by carefully detaching the
origin of the piriformis muscle from the ilium (*beware of injury* to the superi-
or gluteal vessels and superior gluteal nerve!) giving a finger access to check
the sacro-iliac joint from the pelvic interior.

Fig. 12c

1   Gluteus medius muscles
2   Superior gluteal nerve and deep branch of
     superior gluteal vessels
3   Piriformis muscle
4   Superficial branches of
     superior gluteal vessels
5   Inferior gluteal nerve
6   Sacro-tuberous ligament
7   Gluteus maximus muscle

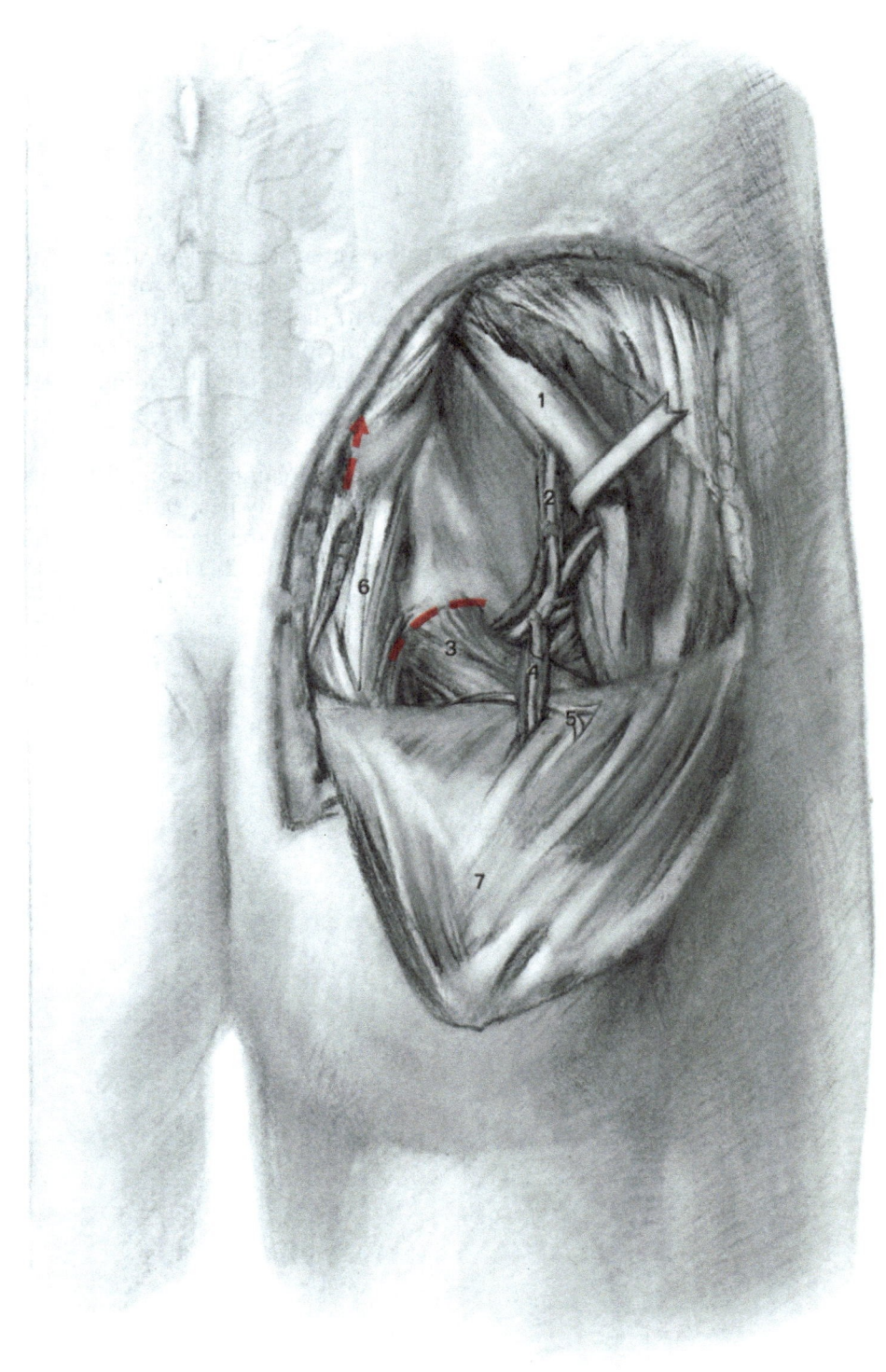

Deep Dissection (continued)

In order to palpate the anterior aspect of the ala or lateral mass of the sacrum from this approach, the rigid fibers of the erector spinae are divided at their insertions on the medial surface of the iliac crest.

In case of a complete disruption of the sacro-iliac joint the ala of the sacrum can be easily reached by inserting the palpating finger through preexisting tears in the tissues.

Internal Fixation

Fixation of the sacro-iliac joint can be accomplished with 2 cancellous bone screws. They are placed in a strictly perpendicular direction to and through the outer surface of the ilium into the ala and the lateral mass of the sacrum. The starting point (shaded area) lies 2–3 fingerwidths lateral to the posterior superior iliac spine and 2 fingerwidths cranial to the greater sciatic notch.

Extension

Practically none.

Closure

The gluteus maximus (and erector spinae if divided) are reapproximated with sutures, and the skin is closed.

Fig. 12d

1   Projection of articular surface
2   Iliac crest
3   Sacro-tuberous ligament
4   Greater sciatic foramen

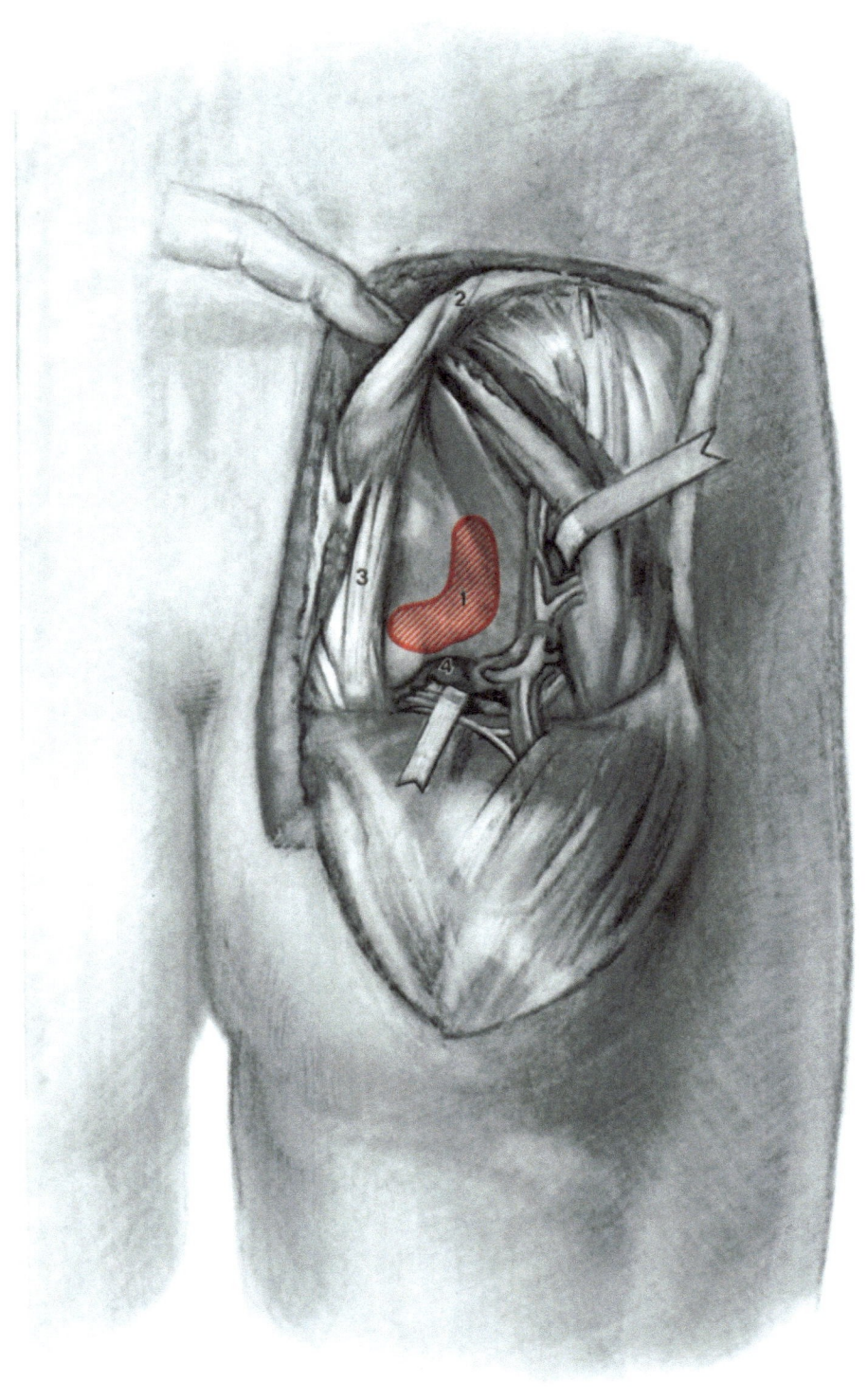

13 Symphysis Pubis and Superior Pubic Rami

Indication
: Rupture of the symphysis pubis, markedly displaced fractures of the anterior pelvic ring.

Positioning/ Draping
: Supine with both anterior superior iliac spines freely accessible.

Skin Incision
: A horizontal incision about 15–20 cm long is placed 1–2 fingerwidths above the *symphysis* (PFANNENSTIEL).

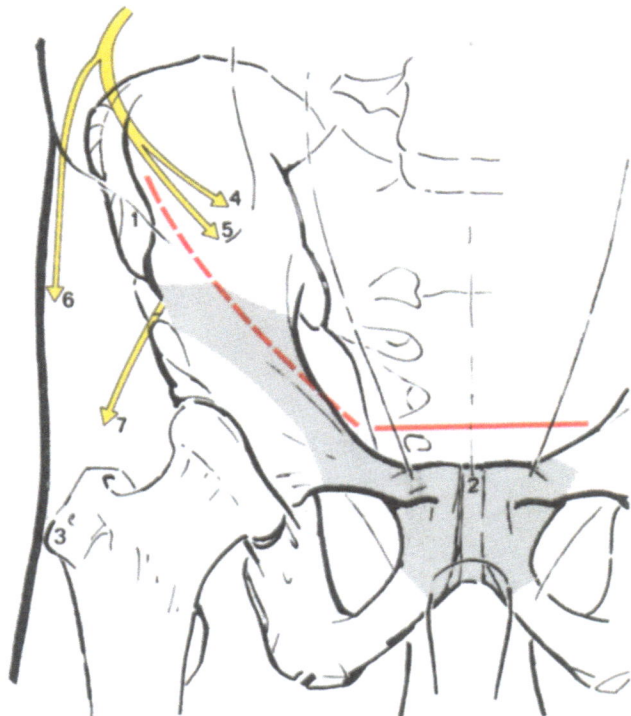

Fig. 13a

1 Anterior superior iliac spine
2 Symphysis pubis
3 Greater trochanter
4 Ilio-hypogastric nerve
5 Ilio-inguinal nerve
6 Lateral cutaneous branch of ilio-hypogastric nerve
7 Lateral cutaneous nerve of thigh

Deep Dissection   The anterior layers of the rectus abdominis sheath (aponeurosis of the obliquus externus and internus muscles), which usually are intact, are divided transversely. Frequently both rectus abdominis muscles will have been stripped from their attachments to the pubic rami. If not, they are surgically released, leaving a narrow strip on the anterior pubic rim.

Internal Fixation   A 2- to 4-hole semitubular plate, DCP, or 3.5 reconstruction plate is applied to the narrow, cranial aspect of both pubic rami. This will enable very long screws to gain a secure hold in the frontal plane.

Fig. 13b

1   Superficial (subcutaneous) inguinal ring
2   Spermatic cord
3   Linea alba

Fig. 13c

1   Rectus abdominis muscles (detached)
2   Pubic crest
3   Preperitoneal space of Retzius

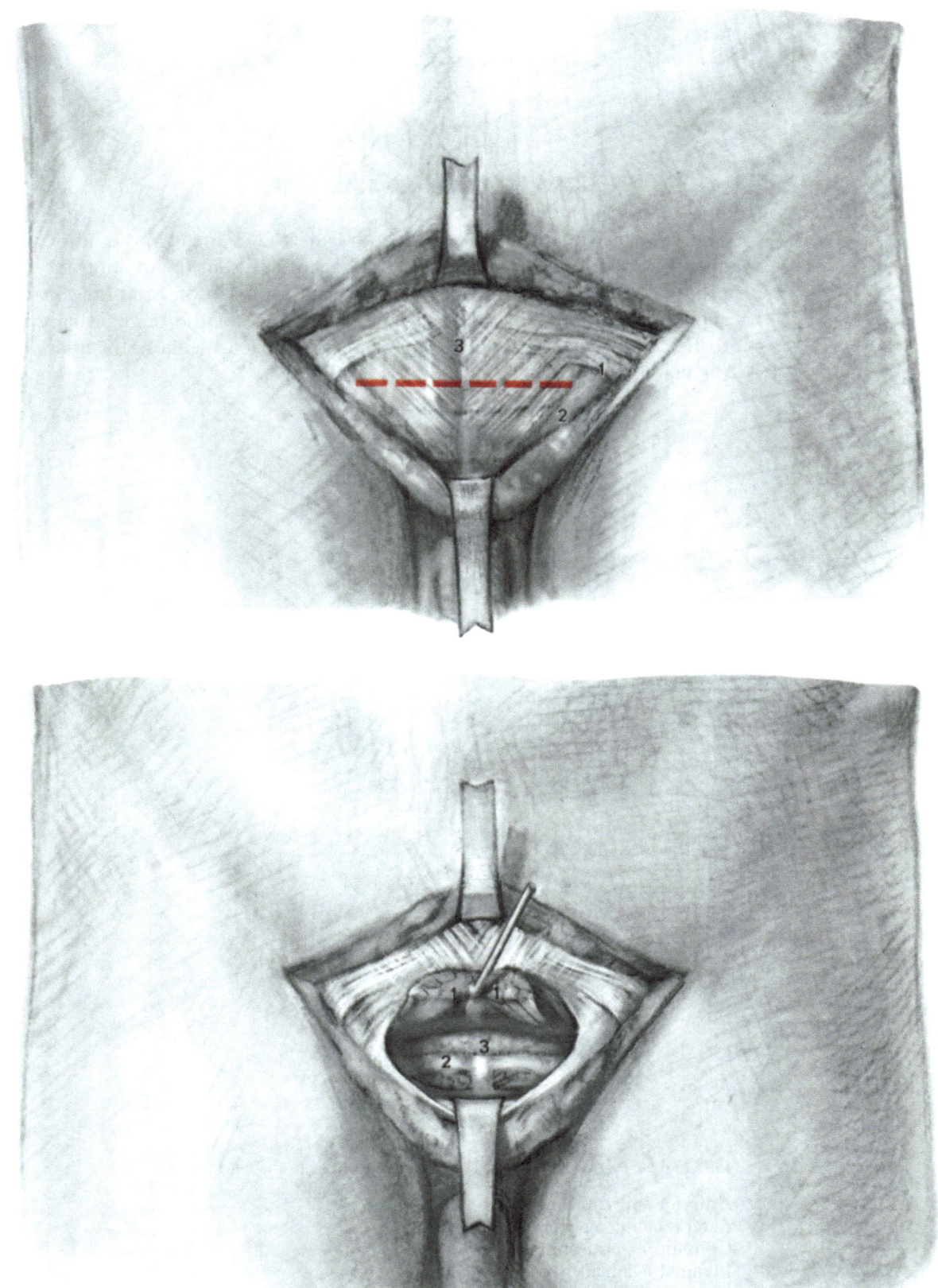

Extension   *In the ilio-inguinal direction*
The skin incision is extended to as much as 2–3 fingerwidths above the anterior superior iliac spine. The lacuna musculorum (containing the iliopsoas muscle and femoral nerve) is opened by osteotomizing the anterior superior iliac spine. The lateral cutaneous nerve of the thigh usually has to be sacrificed. The inguinal ligament is retracted medially together with the iliac fascia.

Fig. 13d

1   Anterior superior iliac spine
2   Lateral cutaneous nerve of thigh
3   Obliquus externus muscle
4   Inguinal ligament
5   Superficial (subcutaneous) inguinal ring
6   Linea alba
7   Incision of Fig. 13b/c

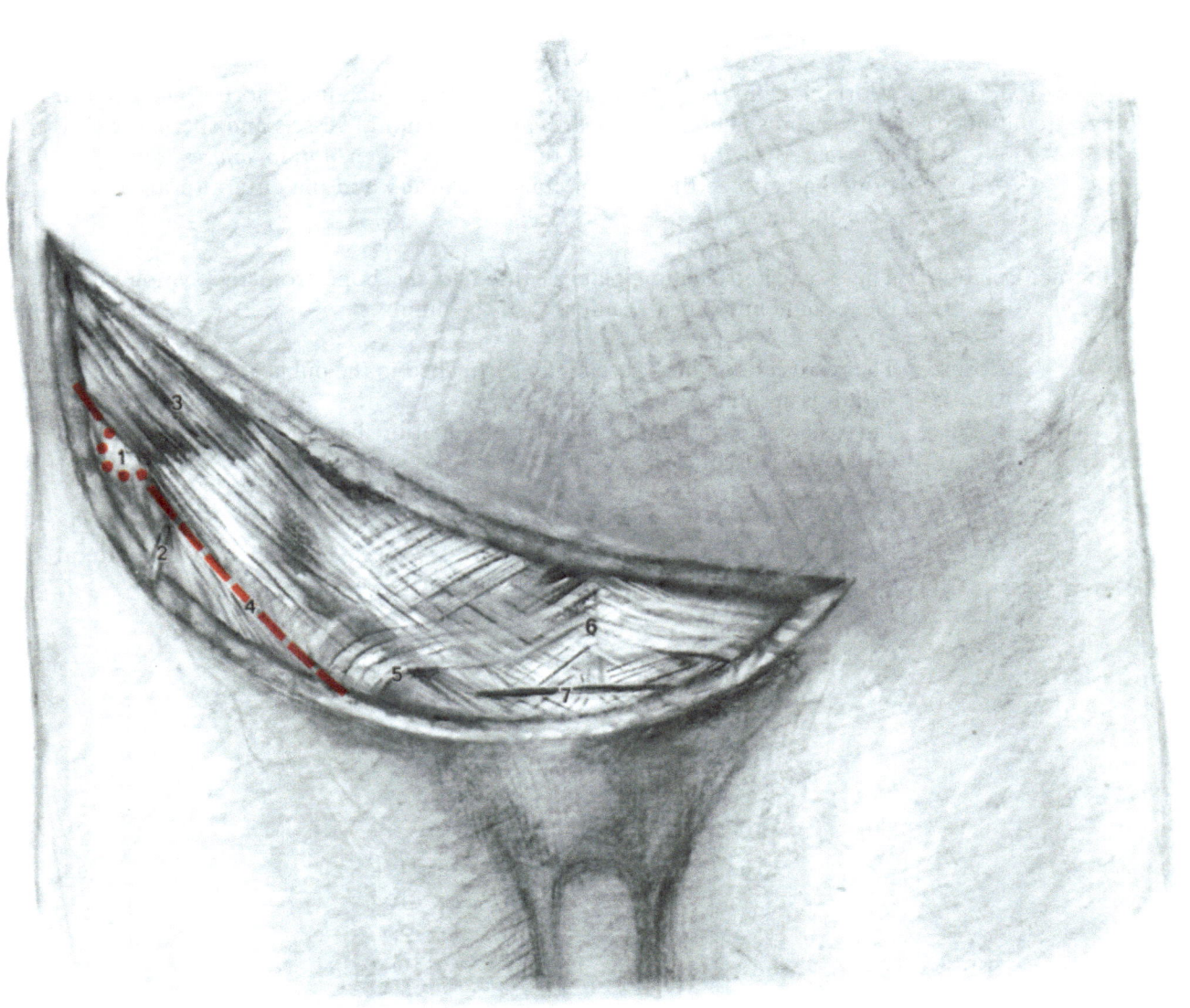

Extension
(continued)

Working from the symphysis and from the lacuna musculorum, the anterior pubic ramus is bluntly exposed, and the femoral vessels and spermatic cord are snared en bloc. At this time care is taken *not to injure* any vessels bridging between the inferior epigastric artery and the obturator artery (the "corona mortis").

Internal Fixation

The 3.5 reconstruction plate is fitted to the bone from the symphysis to the arcuate line (linea terminalis) of the ilium.

Closure

The rectus abdominis muscles are reattached to the pubis, the anterior rectus sheath is sutured.
The osteotomized tip of the anterior superior iliac spine is reattached and the skin is closed.

Hazards

Laterally
Lesions of the spermatic cord, femoral vessels, femoral nerve or lateral cutaneous nerve of the thigh, "corona mortis".

Retrosymphysially
Injury to the urinary bladder or vesico-prostatic venous plexus.

Fig. 13e

1   Anterior superior iliac spine (after osteotomy)
2   Lateral muscles of abdominal wall
3   Inguinal ligament
4   Iliac fossa with iliac muscle
5   Psoas major muscle
6   Superior ramus of pubis
7   Spermactic cord
8   Lateral cutaneous nerve of thigh
9   Rectus abdominis muscles (detached)

14 Hip Joint: Posterior Approach

Indications   Femoral head arthroplasty, isolated fractures of the posterior acetabular rim.

Positioning/
Draping   Lateral on a vacuum mattress with the leg freely mobile. The draping leaves most of the buttock uncovered.

Skin Incision   Curves anteriorly from the *posterior superior iliac* spine across the *greater trochanter,* ending about 3–4 cm distal to the trochanter.

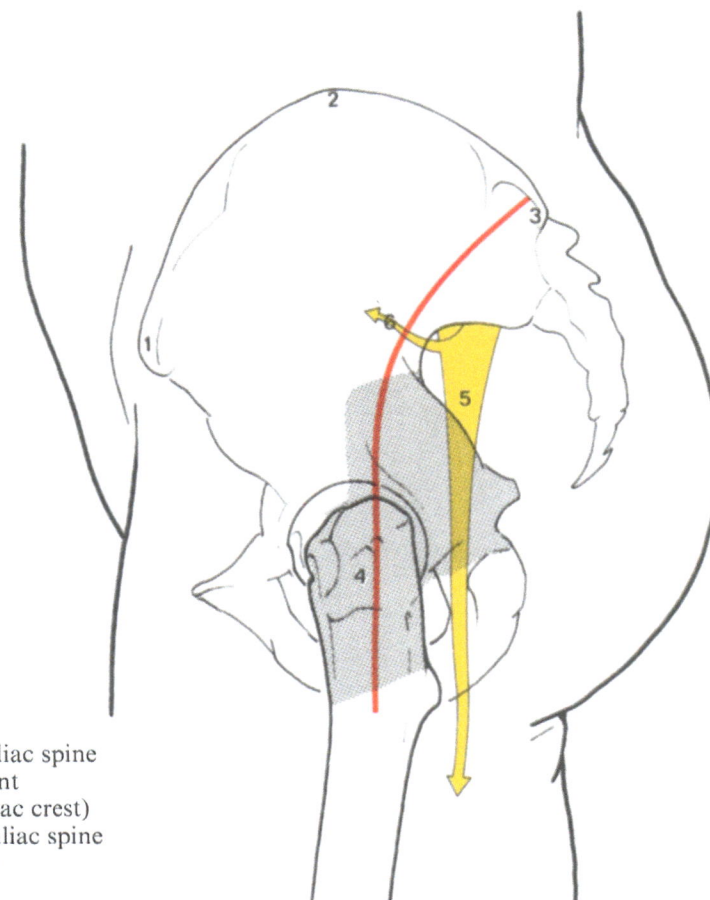

Fig. 14a

1   Anterior superior iliac spine
2   "Supracristal" point
    (highest point of iliac crest)
3   Posterior superior iliac spine
4   Greater trochanter
5   Sciatic nerve
6   Superior gluteal nerve

Deep Dissection   The ilio-tibial tract is split over the greater trochanter, and the incision is extended in line with the fibers of the gluteus maximus.

Fig. 14b

1   Greater trochanter
2   Ilio-tibial tract
3   Gluteus maximus muscle

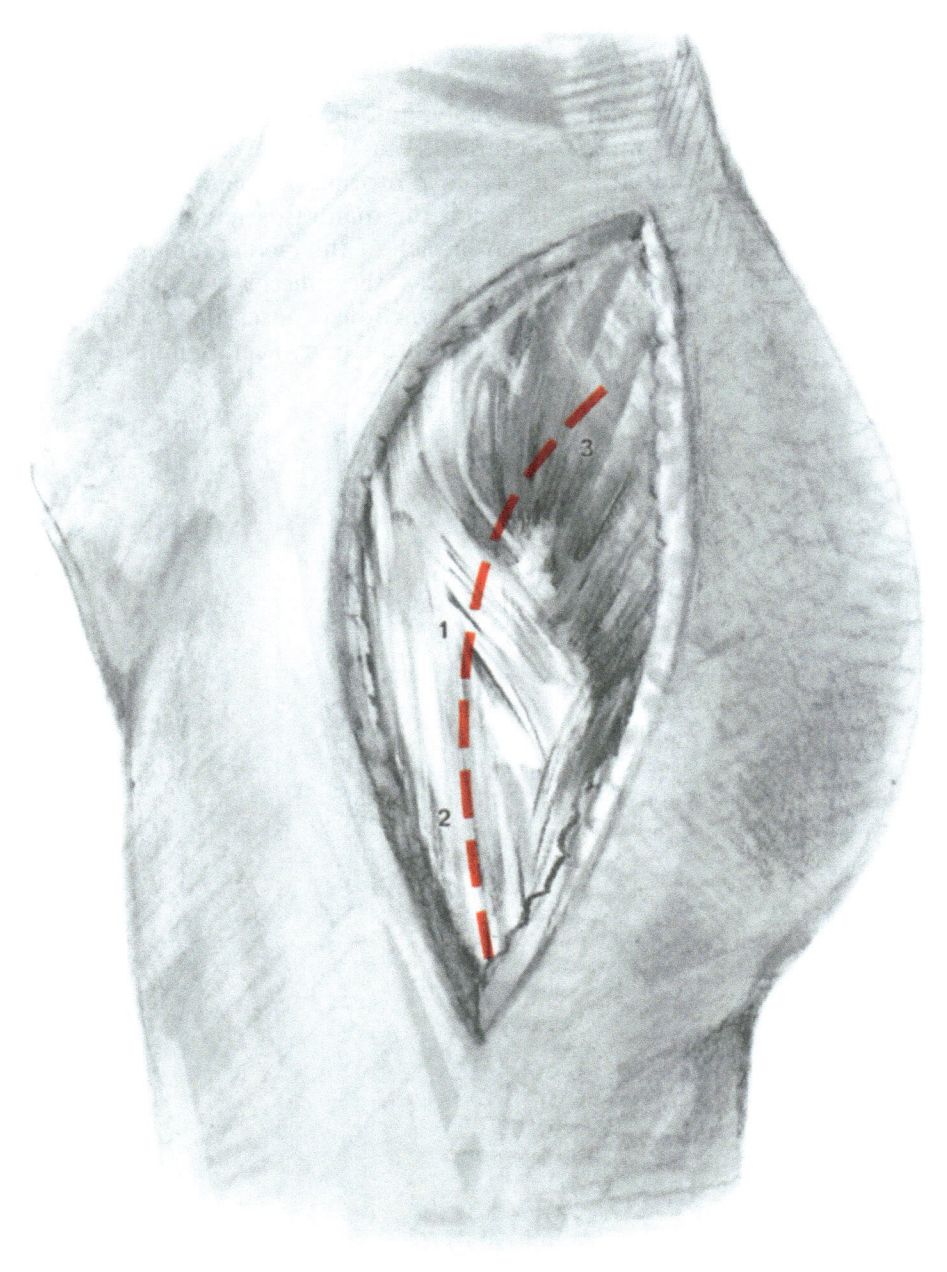

Deep Dissection (continued)

The femur is rotated internally, and the small external rotators (piriformis, gemelli and internal obturator muscles) are divided close to the greater trochanter. Retraction of the piriformis is maintained with stay sutures. It may be necessary to notch the quadratus femoris muscle at its cranial border and ligate the deep branch of the medial femoral circumflex artery. When reflected postero-medially, the muscles will cover and protect the sciatic nerve. Their careful retraction exposes the posterior joint capsule and the posterior rim and roof of the bony acetabulum.

Fig. 14c

1 Greater trochanter
2 Quadratus femoris muscle
3 "Triceps coxae" (gemelli and internal obturator muscles)
4 Piriformis muscle
5 Gluteus medius muscle
6 Sciatic nerve
7 Gluteus maximus muscle
8 Superficial branch of superior gluteal vessels
9 Inferior gluteal nerve and vessels

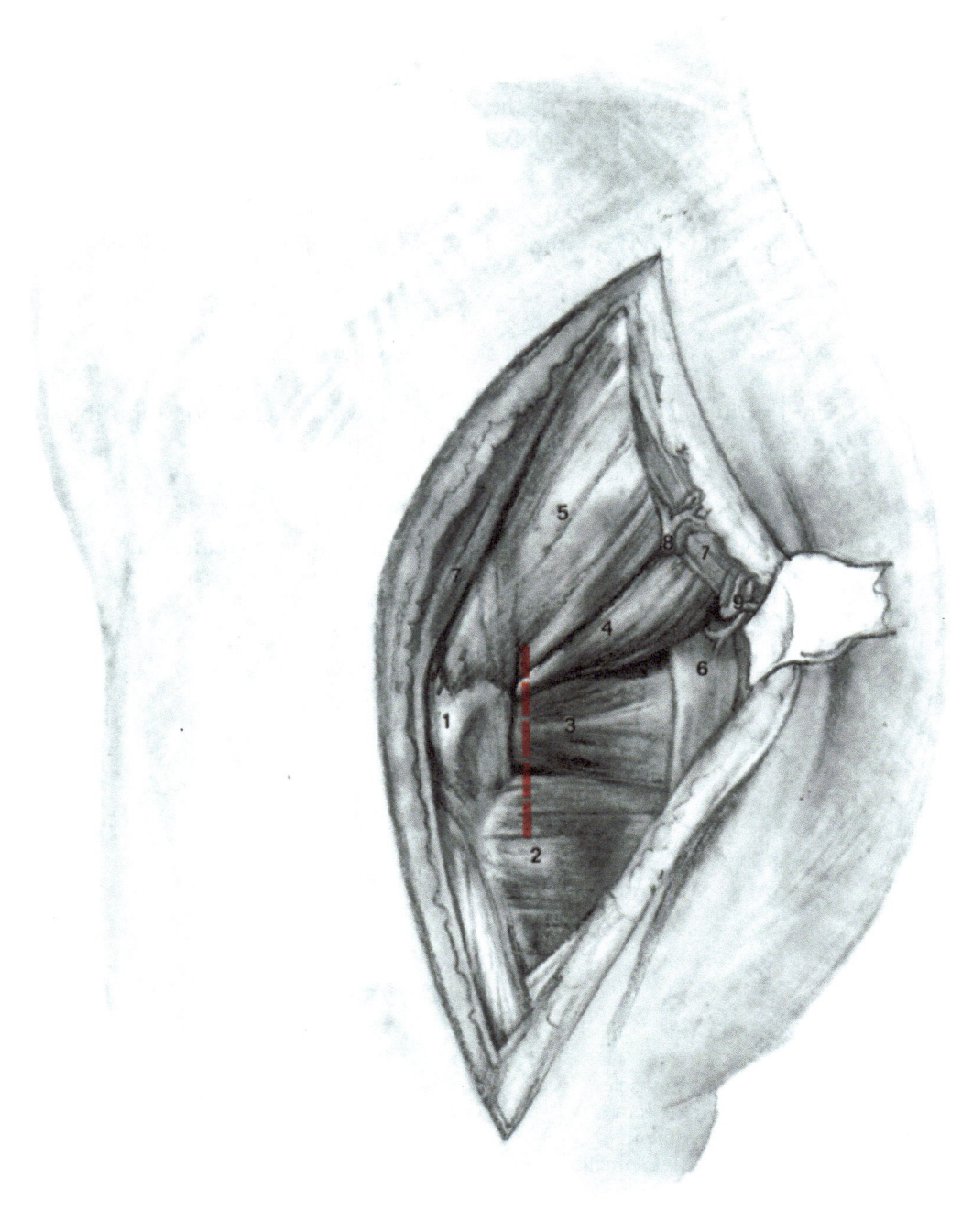

Extension   *Proximally*
Notching the insertions of the "small" gluteal muscles on the greater trochanter increases proximal exposure to some degree.

Posteriorly
Osteotomy of the ischial spine enables the index finger to palpate from
the back the body of the ischium or part of the "anterior column" of
the acetabulum.

Distally
Along the axis of the femoral shaft (see Chap. 18).

Closure   The piriformis (and "triceps coxae") are reattached, the ilio-tibial tract
is reapproximated and sutured, and the skin is closed.

Hazards   – Retractor pressure on the sciatic nerve and inferior gluteal nerve.

– Injury of the superior gluteal vessels and superior gluteal nerve at their
site of emergence from the suprapiriform foramen.

Fig. 14d

1   Gluteus medius muscle
2   Piriformis muscle (cut)
3   Sacrospinal ligament
4   "Triceps coxae" (gemelli and internal obturator muscles, (cut)
5   Quadratus femoris muscle (cut)
6   Vastus lateralis muscle
7   Deep branch of circumflex femoral artery
8   Inferior gluteal nerve and inferior gluteal vessels
9   Superior gluteal vessels
10 Sciatic nerve
11 Ischial spine
12 Joint capsule

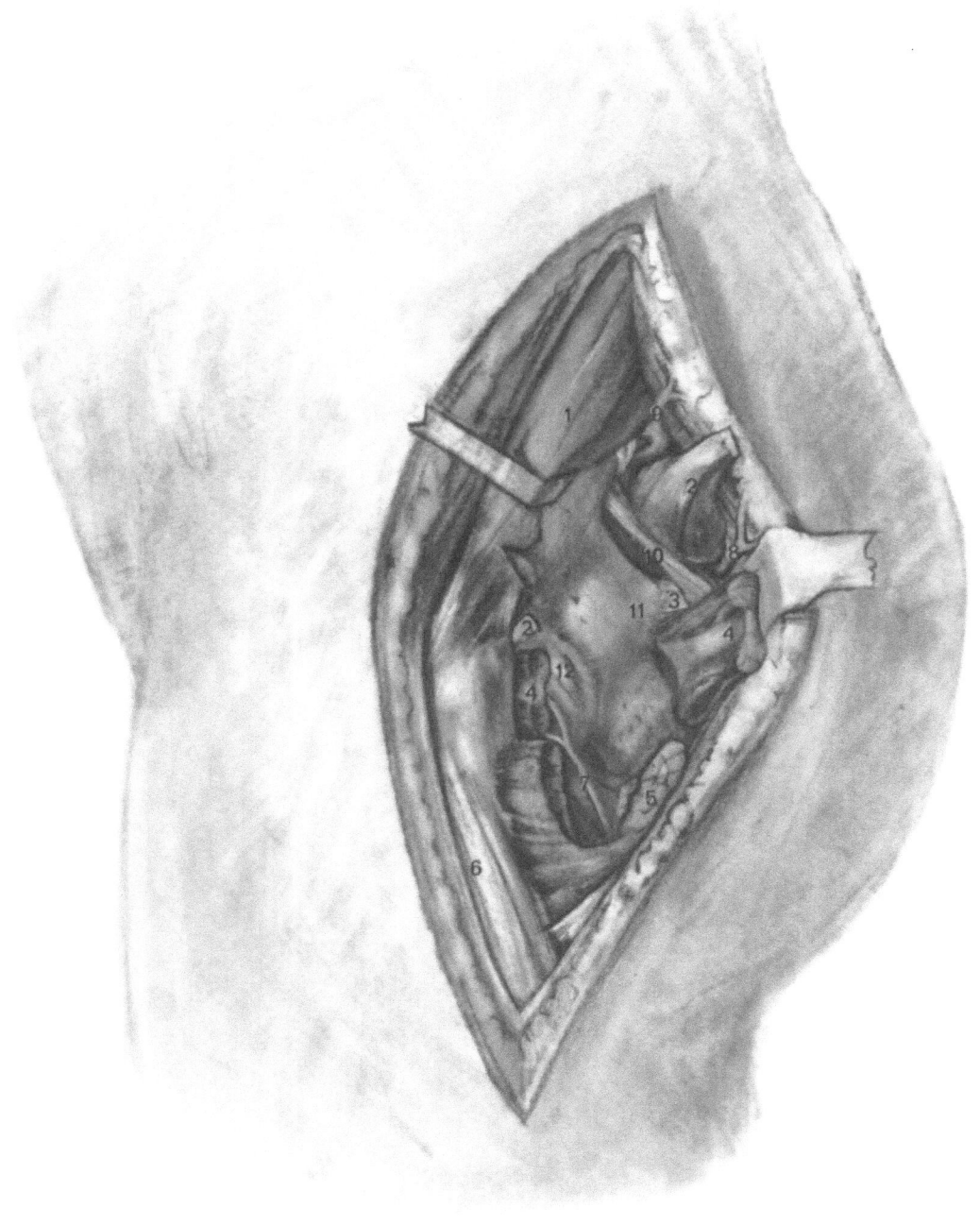

15 Hip Joint: Postero-lateral Approach

Indications   Acetabular fractures, especially of the roof and posterior column.

Advantage   Gives broad exposure of the posterior acetabular wall and lower half of the wing of the ilium without endangering the nerve supply to the three gluteal muscles.

Positioning/   Strictly lateral on a vacuum mattress with the leg draped such that it can
Draping   be freely moved. The area is draped free to the symphysis pubis *anteriorly*, to the rear portion of the buttock *posteriorly* (parallel to the anal cleft), to the costal margin *proximally*, and to the midthigh *distally*.

Skin Incision   With the hip joint extended, the incision starts at the *highest point of the iliac crest* and is carried past and slightly posterior to the *greater trochanter*, aiming for the *lateral femoral condyle*. It terminates no more than 10–15 cm distal to the greater trochanter.

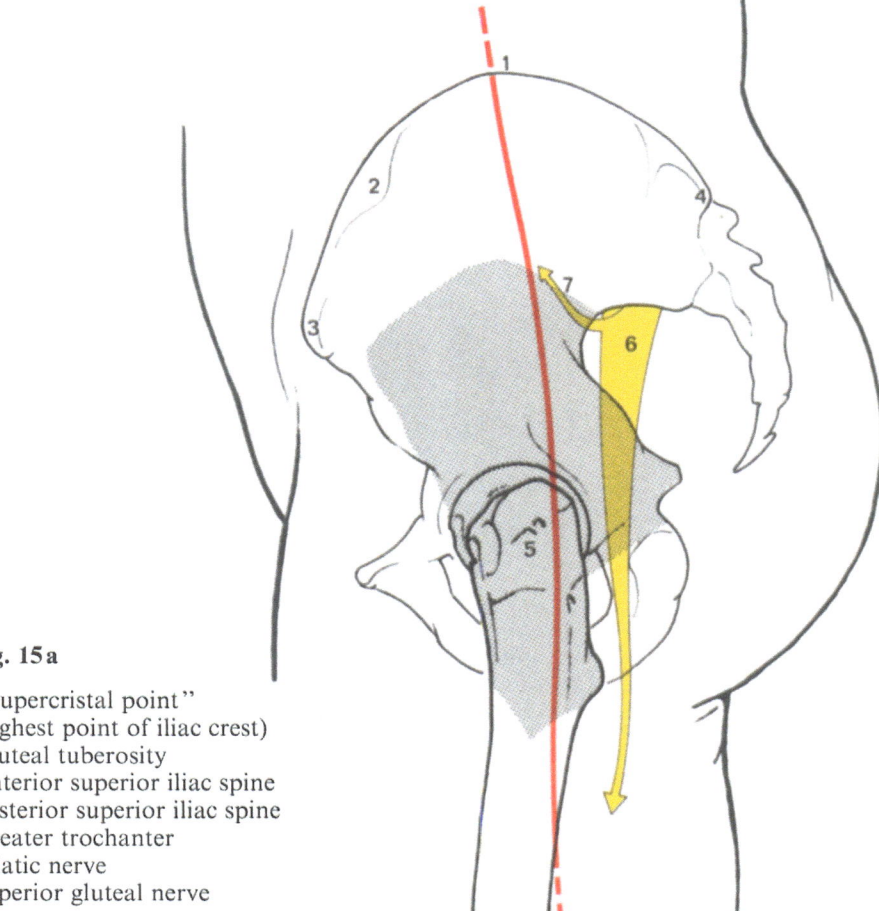

Fig. 15 a

1   "Supercristal point"
     (highest point of iliac crest)
2   Gluteal tuberosity
3   Anterior superior iliac spine
4   Posterior superior iliac spine
5   Greater trochanter
6   Sciatic nerve
7   Superior gluteal nerve

Deep Dissection

An incision is made through the fascia covering gluteus medius muscle and along the fleshy anterior border of gluteus maximus, continuing into the ilio-tibial tract.

The ilio-tibial tract is released from the iliac crest, working forward along the crest until its most lateral point, the gluteal tuberosity, is reached.

Fig. 15b

1 Greater trochanter
2 "Supracristal point" (see 15a/1)
3 Ilio-tibial tract
4 Tensor fasciae latae muscle

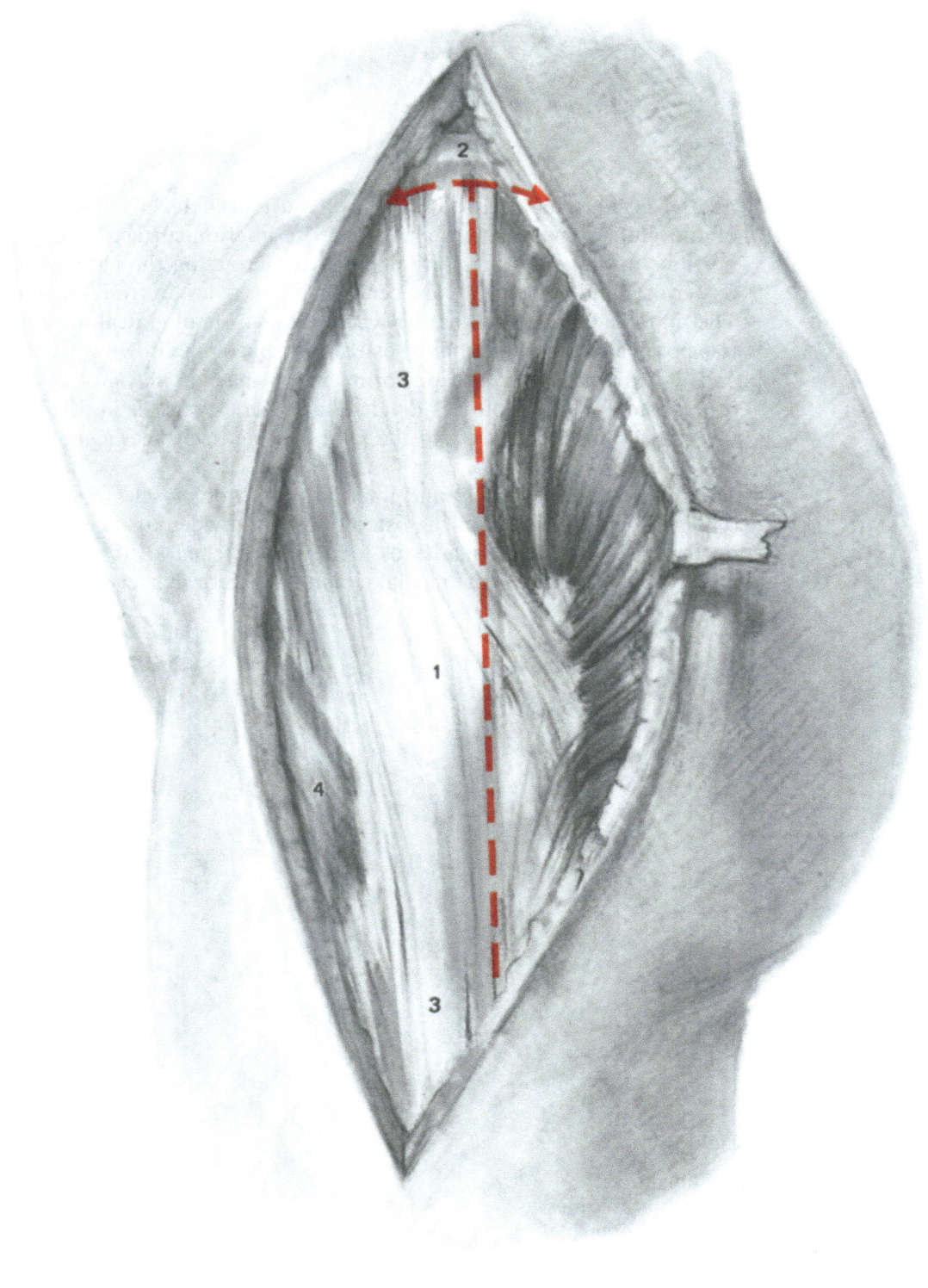

Deep Dissection
(continued)

As the tract is reflected anteriorly, it must be detached from the fibers of the gluteus medius that arise from its inferior surface. The posterior portions of gluteus medius are exposed by retracting the gluteus maximus posteriorly and detaching a few centimeters of its origin from the iliac crest. The tip of the greater trochanter where the two small gluteal muscles insert is exposed and cut by an oblique osteotomy that slopes downward anteriorly and upward medially. The piriform tendon may or may not be included in the osteotomy; if not, it is divided later. An elevator is inserted between the gluteal muscle pedicle and the joint capsule to protect against inadvertantly opening the joint.

The triceps coxae (gemelli, and internal obturator muscles) and piriformis (unless already included in the osteotomy) are divided at their attachments on the trochanter. The quadratus femoris is notched at its cranial border, and the deep branch of the medial circumflex femoral artery is ligated.

Fig. 15c

1   Ilio-tibial tract
2   Gluteus medius muscle
3   Superficial branch of superior gluteal vessels
4   Piriformis muscle
5   Sciatic nerve
6   Inferior gluteal vessels
7   Gluteus maximus muscle
8   Vastus lateralis muscle
9   "Triceps coxae" (gemelli and internal obturator muscles)
10   Quadratus femoris muscle
11   Tendon of gluteus maximus muscle

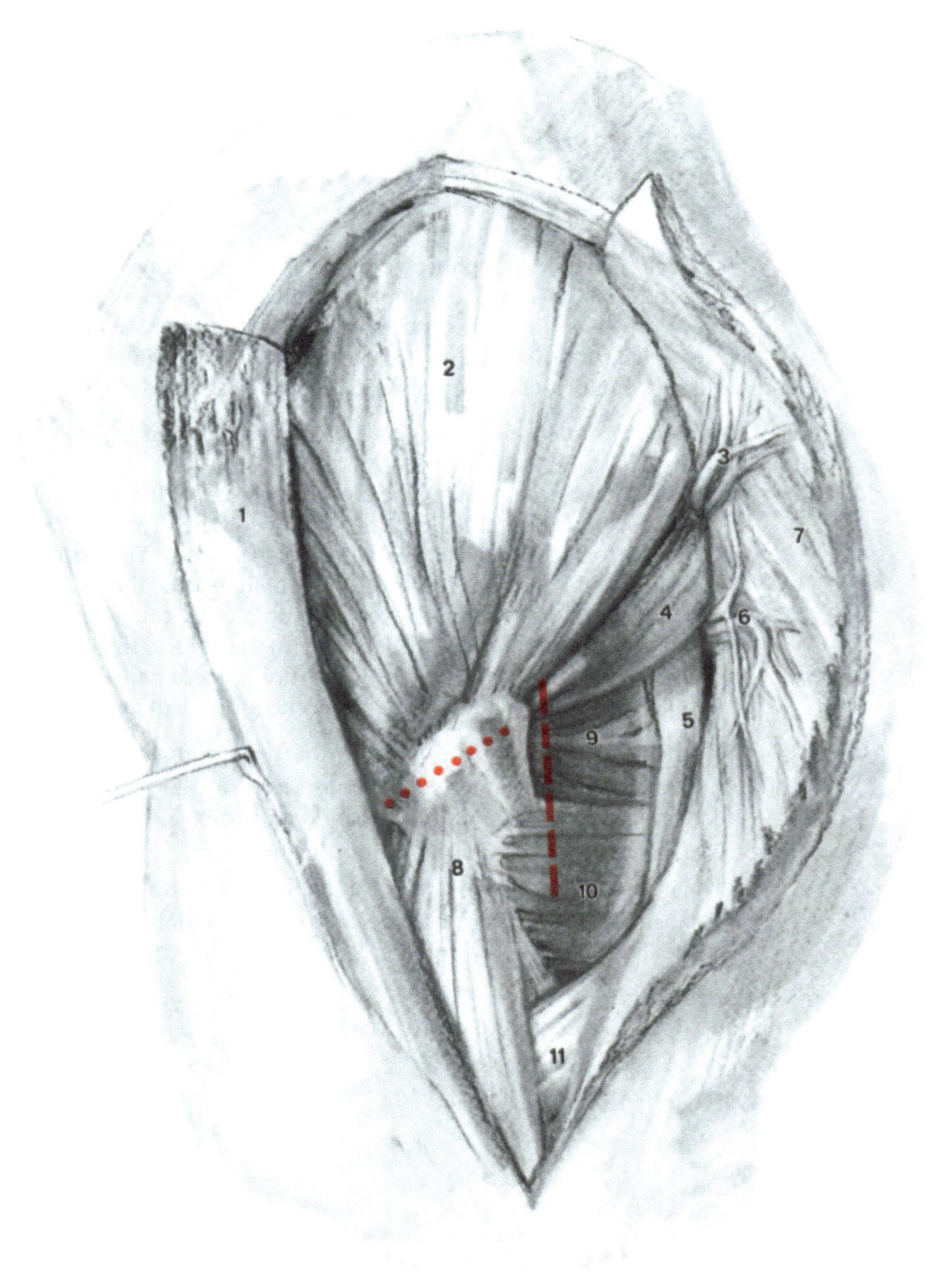

Deep Dissection (continued)   The trochanter with its gluteal attachment is carefully reflected cranially, taking care *not to injure* the superior gluteal vessels and superior gluteal nerve; this exposes the entire acetabular roof and the base of the iliac wing. The external rotators are retracted posteriorly from the joint capsule until the posterior column is exposed; this will also retract and protect the sciatic nerve.

Extension   *To palpate the anterior column or the iliac fossa*, the skin incision is extended cranially.

When the anterior part of the iliac crest is exposed, either it is osteotomized, or the muscular abdominal wall together with the iliopsoas muscle is retracted anteromedially from the wing of the ilium. The palpating index finger can thereby reach as far as the arcuate line of the ilium (linea terminalis) of the anterior column ("over the top") and can assist in reducing the fracture or guiding the drill bit. With the index finger of the other hand in the greater sciatic notch, the entire posteromedial contour of the acetabular rim (ischial body) may be checked for adequate reduction.

Closure   1. The tendons of the piriformis and triceps coxae are sutured.
2. The tip of the trochanter with the insertions of the small gluteal muscles is reattached with screws or a tension band-wire.
3. The ilio-tibial tract is reattached to the iliac crest.
4. The gluteus maximus is sutured to the iliac crest and ilio-tibial tract.
5. The osteotomized anterior pelvic crest or the abdominal wall is readapted to the iliac crest.

Hazards   *Posterior side*
Injury to sciatic nerve and neurovascular bundle of the small gluteal muscles.

Anterior side
Nerve to tensor fasciae latae.

Fig. 15d

1   Rectus femoris muscle, reflected head
2   Nerve of tensor fasciae latae
3   Gluteus medius and minimums muscles (reflected)
4   Greater trochanter (after osteotomy)
5   Tendon of piriformis muscle (cut)
6   Sciatic nerve
7   Superficial branch of superior gluteal vessels
8   Origin of gluteus minimus
9   Superior gluteal nerve
10   "Triceps coxae"
     (gemelli and internal obturator muscles)
11   Quadratus femoris muscle
12   Vastus lateralis muscle
13   Tensor fasciae latae muscle

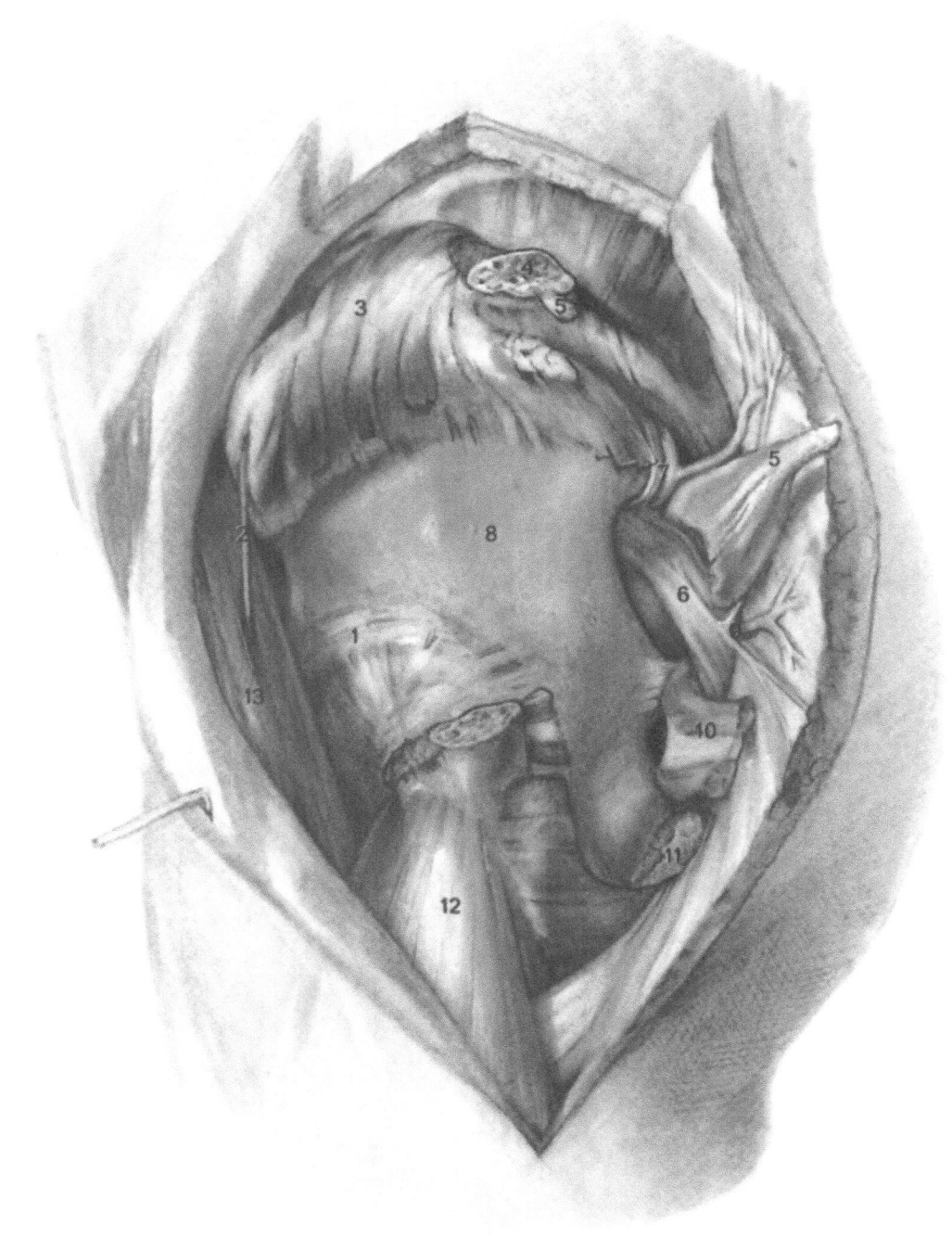

16 Hip Joint: Anterior Ilio-inguinal Approach

Indication
Fractures of the anterior column, superior pubic ramus, or upper ilium.

Positioning/ Draping
Supine with the hip slightly elevated (vacuum mattress) and the leg draped so that it can be freely moved.

Skin Incision
Starting at the highest point on the *iliac crest*, the incision passes along the crest to the *anterior superior iliac spine* and from there curves medially to a point 2 fingerwidths above the *symphysis pubis*.

Fig. 16a

1 Anterior superior iliac spine
2 Symphysis pubis
3 Greater trochanter
4 Ilio-hypogastric nerve
5 Ilio-inguinal nerve
6 Lateral cutaneous branch of ilio-hypogastric nerve
7 Lateral cutaneous nerve of thigh

Deep Dissection   The iliacus muscle is released, together with the abdominal wall, from the iliac crest and the iliac fossa down to the anterior superior iliac spine, leaving a narrow strip of tissue. At the spine the attachment of the inguinal ligament is cut or osteotomized, which often entails division of the lateral cutaneous nerve of the thigh.

Fig. 16b

1   Anterior superior iliac spine
2   Obliquus externus muscle
3   Lateral cutaneous nerve of the thigh
4   Superficial epigastric vein
5   Ilio-inguinal nerve
6   Spermatic cord

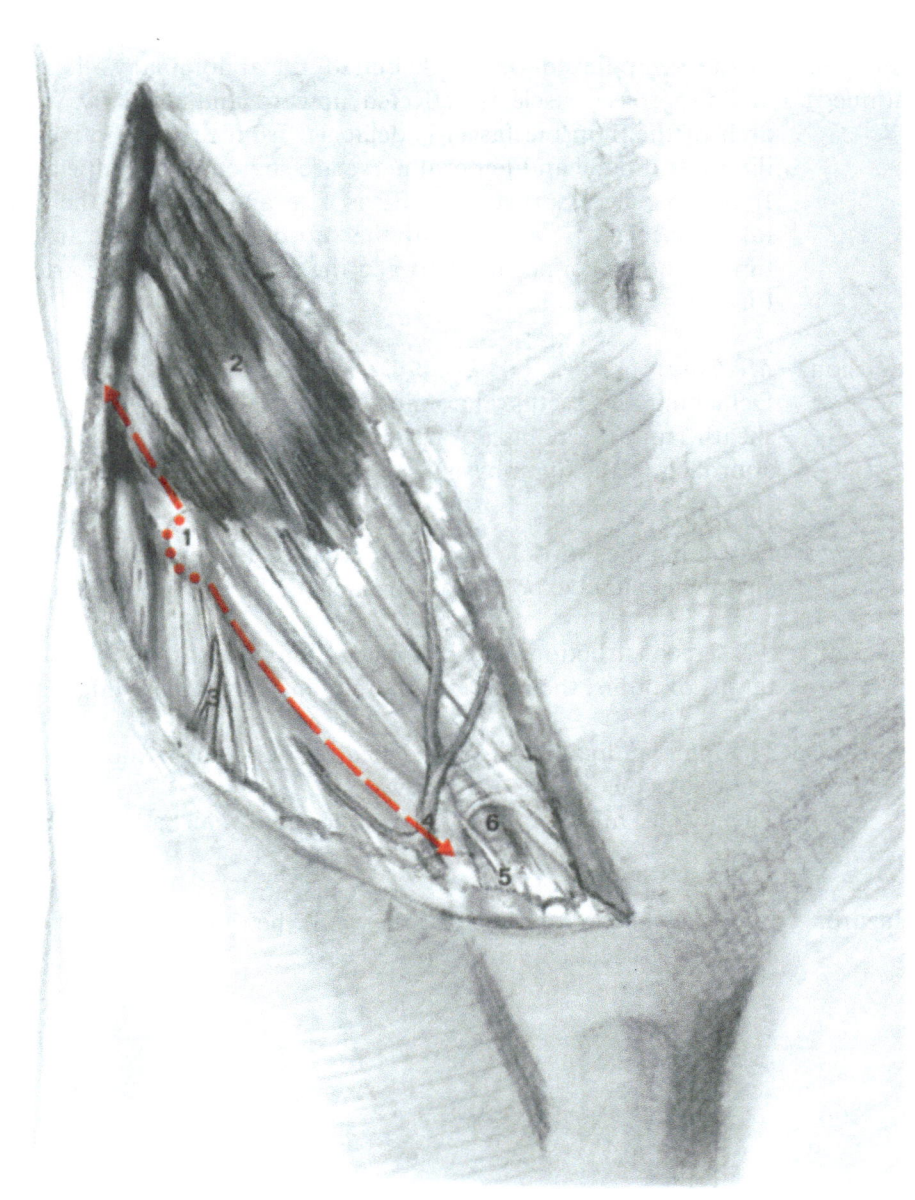

Deep Dissection (continued)

With the hip flexed, the whole bulk of the abdominal wall, inguinal ligament and iliopsoas muscle is reflected upward and medially. The iliopectineal arch of the iliopsoic fascia is detached from the iliopectineal tubercle. The iliopsoas muscle and femoral nerve are snared with a tape and retracted.

It may be necessary to bluntly isolate and mobilize the femoral vessels, taking special care *not to injure* the "corona mortis," the arterial (and sometimes venous) connection between the inferior epigastric artery and obturator artery.

Extension

Medially
Detaching the rectus abdominis muscle and anterior layer of the rectus sheath from the pubis affords access to the pre- and paravesical space (cf. approach to the Symphysis).

Posteriorly
Detaching the iliopsoic fascia from the arcuate line of the ilium (linea terminalis), will extend exposure as far back as the sacro-iliac joint.

Internal Fixation

A 3.5 reconstruction plate is applied along the narrow aspect of the superior ramus of pubis, the anterior column, and the arcuate line.

Closure

The inguinal ligament is reattached to the anterior superior iliac spine with a transosseous suture or a screw.
The iliacus muscle and lateral abdominal wall is reattached to the iliac crest.

Hazards

Injury to the lateral cutaneous nerve of the thigh or to the "corona mortis" artery or associated veins.

Fig. 16c

1   Anterior superior iliac spine (after osteotomy)
2   Iliac fossa
3   Lateral muscles of abdominal wall
4   Iliac muscles
5   Fascia lata (cut)
6   Lateral cutaneous nerve of the thigh (cut)
7   Deep circumflex iliac artery (cut)
8   Spermatic cord with ilio-inguinal nerve

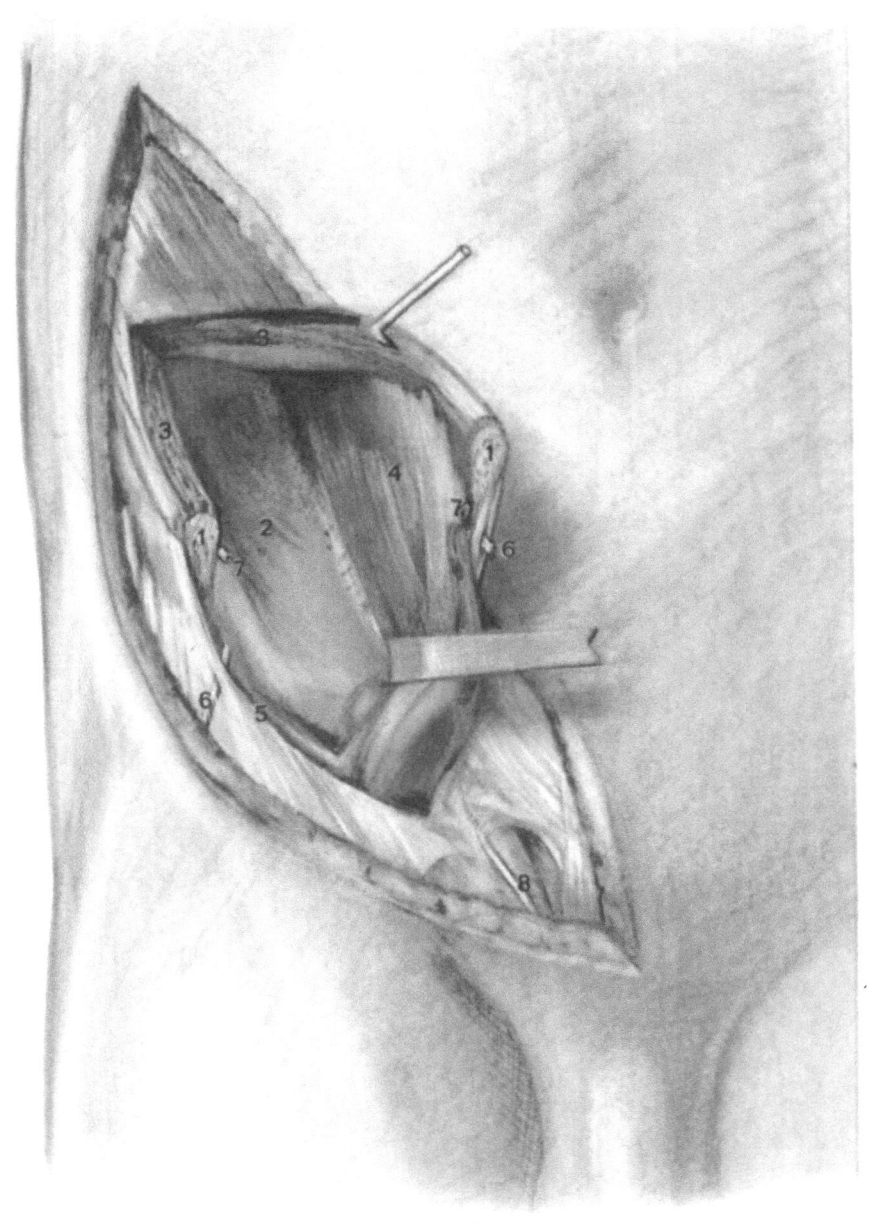

17 Proximal Femur: Lateral Approach

Indications	Femoral neck fractures (including head prostheses); per-, inter-, and subtrochanteric fractures of the femur.
Positioning/ Draping	Supine with some padding beneath the buttock; the leg is draped so that is can be freely moved; or a traction-table is used.
Skin Incision	The incision starts 1–2 fingerwidths proximal to the palpable *greater trochanter* and is carried straight toward the *lateral femoral condyle* for a distance of 10–15 cm.

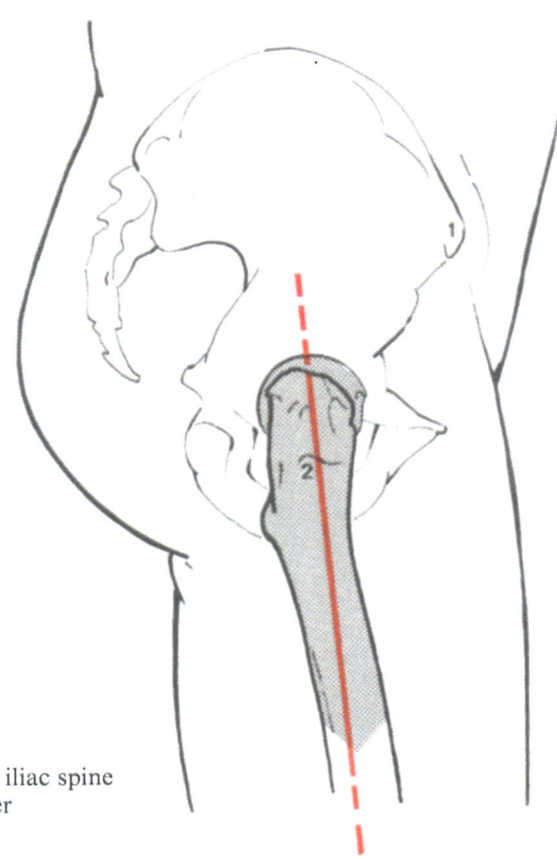

Fig. 17a

1   Anterior superior iliac spine
2   Greater trochanter

Deep Dissection   The ilio-tibial tract is split longitudinally over the greater trochanter, carrying the incision distalward in line with the fibers. The subfascial trochanteric bursa is removed.

Fig. 17b

1   Tensor fasciae latae
2   Ilio-tibial tract
3   Greater trochanter

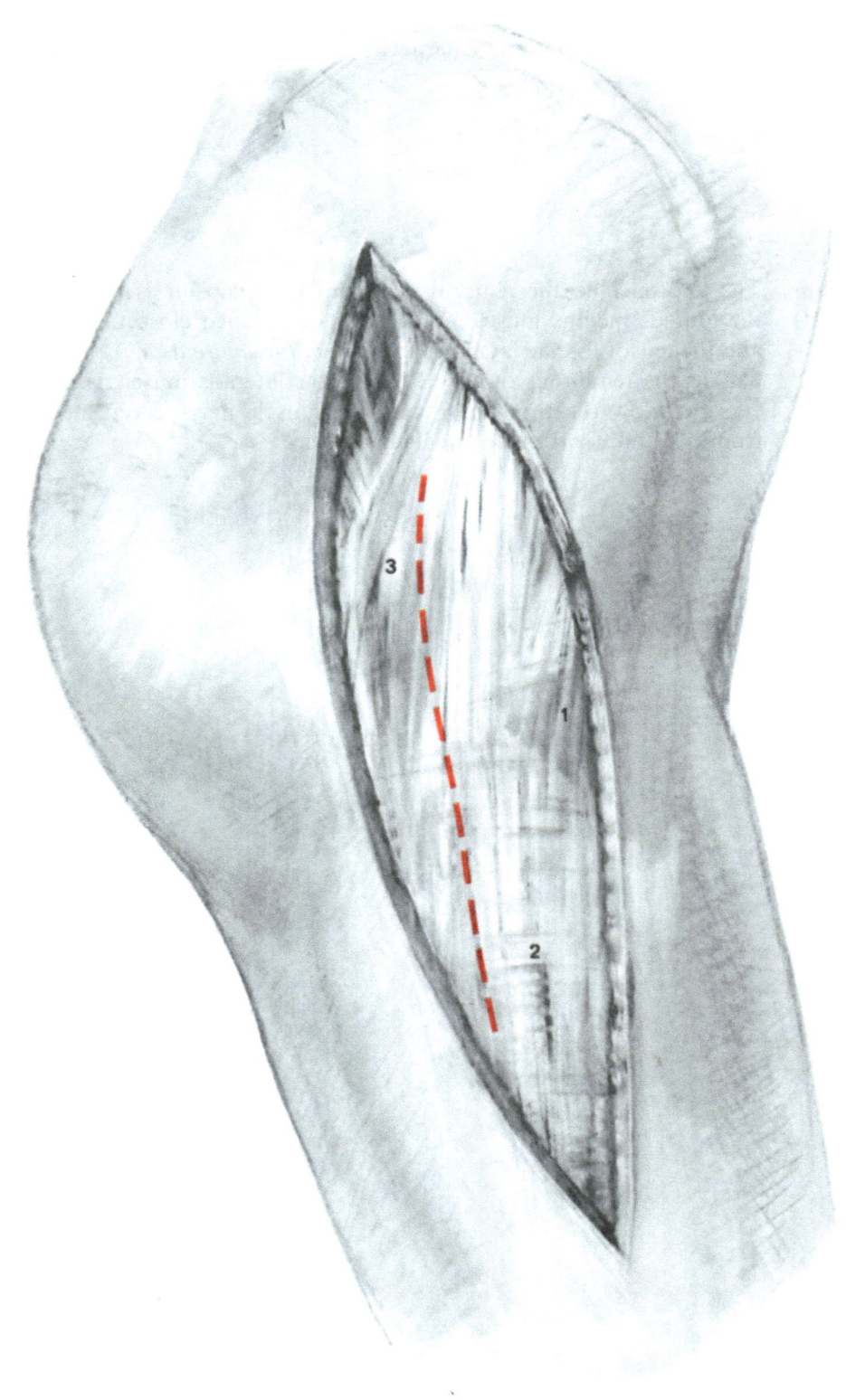

Deep Dissection   The approach to the femoral neck and joint capsule is prepared between
(continued)   the gluteus medius muscle, which can be notched close to the trochanter,
                  and the tensor fasciae latae (taking care *not to injure its nerve*). The posterior
                  half of the tendinous origin of the vastus lateralis muscle is incised trans-
                  versely below the trochanter, and the muscle mass is reflected anteriorly
                  from the bone, starting at the linea aspera.

Fig. 17c

1   Greater trochanter
2   Vastus lateralis muscle
3   Fascia lata
4   Gluteus medius muscle
5   Lateral circumflex femoral vessels
6   Frohse band

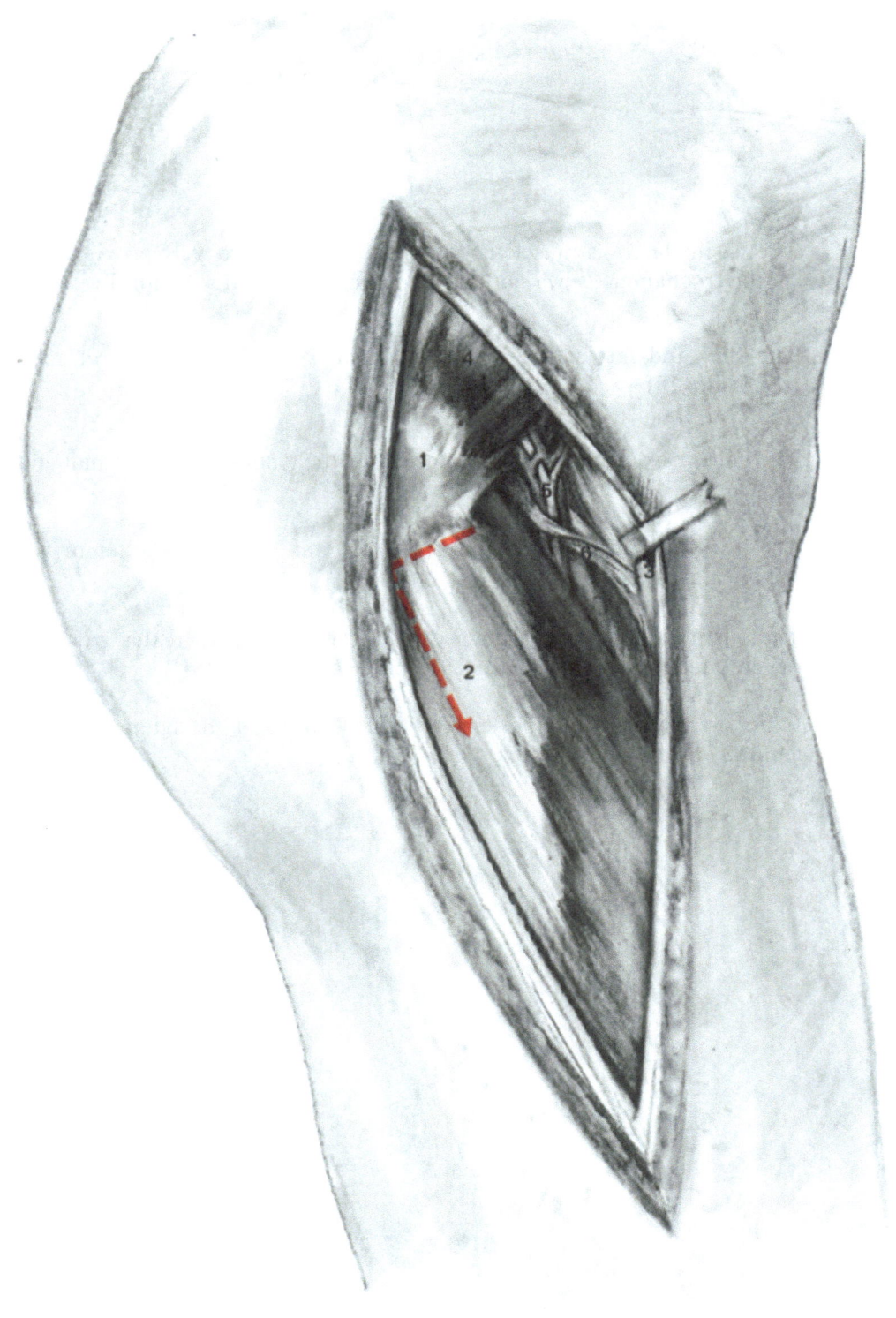

Extension   *Medially*
             The vastus lateralis muscle is completely released from its proximal origin on the trochanter (*beware of injury* to lateral circumflex femoral vessels).

             Distally
             The skin and fascial incision is extended to the lateral femoral condyle (see Chap. 18).

             Proximally
             The greater trochanter is osteotomized, or the tendons of both small gluteal muscles are divided.

Internal Fixation   The side plate is applied to the postero-lateral surface of the femur, along the linea aspera.

Closure   The vastus lateralis is reattached to the base of the greater trochanter; the cut gluteus medius and the ilio-tibial tract are sutured.

Hazards   Injury to the nerve of the tensor fasciae latae and the lateral circumflex femoral vessels.

Fig. 17d

1   Gluteus medius muscle
2   Capsule of hip joint
3   Nerve to tensor fasciae latae
4   Iliopsoas muscle
5   Lateral circumflex femoral vessels
6   Vastus intermedius muscle
7   Greater trochanter
8   Vastus lateralis muscle (reflected)
9   Perforating artery, posterior vastal branch

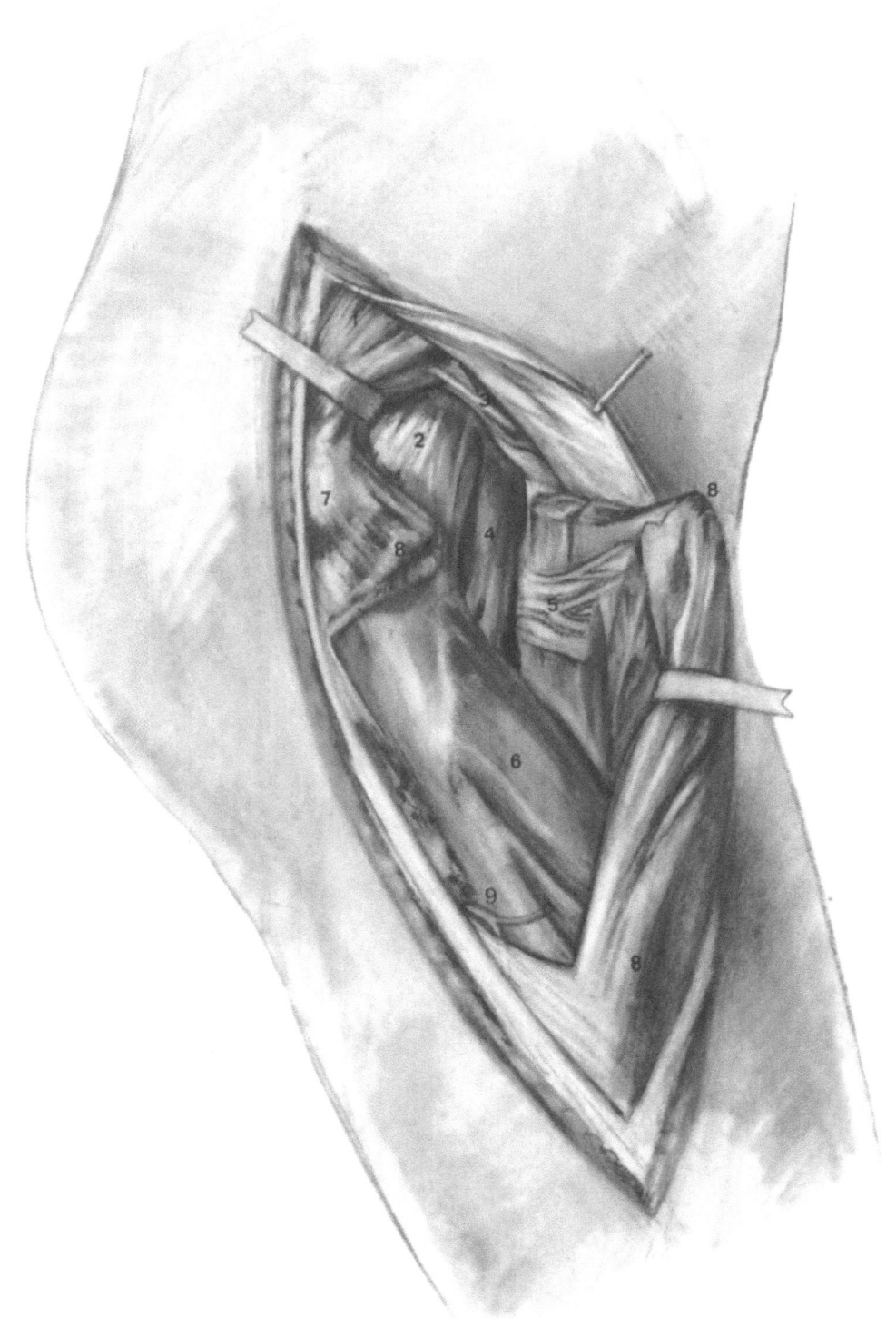

18 Femoral Shaft: Lateral Approach

Indications   Open reduction and plating femoral shaft fractures, with or without metaphyseal extension.

Positioning/ Draping   Lateral on a vacuum mattress; the leg is draped so that is can be freely moved; or a traction-table is used.

Skin Incision   It is made on a straight line connecting the *greater trochanter and lateral femoral condyle*. Its length depends upon the location and extent of the fracture.

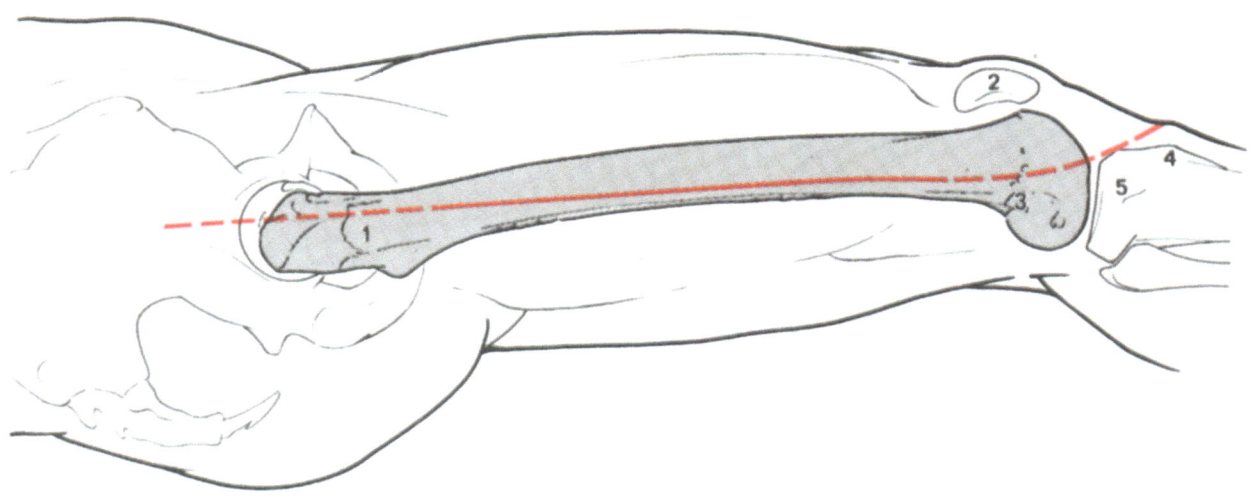

Fig. 18a

1   Greater trochanter
2   Patella
3   Lateral condyle of femur
4   Tuberosity of tibia
5   Tubercle of GERDY

Deep Dissection   The ilio-tibial tract is split in line with its fibers. Using a combination of sharp and blunt dissection, the vastus lateralis is separated from the ilio-tibial tract and from its continuation to the bone (lateral intermuscular septum), back to the the linea aspera.

Fig. 18b

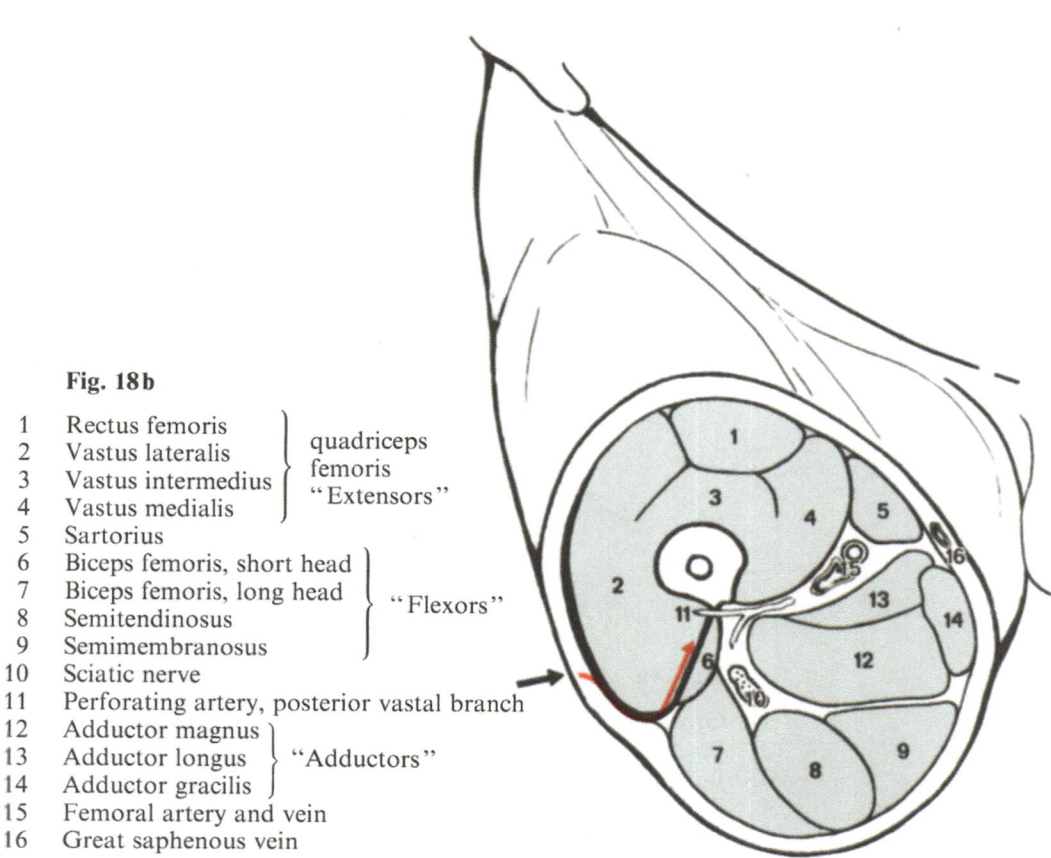

1   Rectus femoris   ⎫
2   Vastus lateralis   ⎪ quadriceps
3   Vastus intermedius   ⎬ femoris
4   Vastus medialis   ⎭ "Extensors"
5   Sartorius
6   Biceps femoris, short head ⎫
7   Biceps femoris, long head   ⎪
8   Semitendinosus   ⎬ "Flexors"
9   Semimembranosus   ⎭
10   Sciatic nerve
11   Perforating artery, posterior vastal branch
12   Adductor magnus   ⎫
13   Adductor longus   ⎬ "Adductors"
14   Adductor gracilis ⎭
15   Femoral artery and vein
16   Great saphenous vein

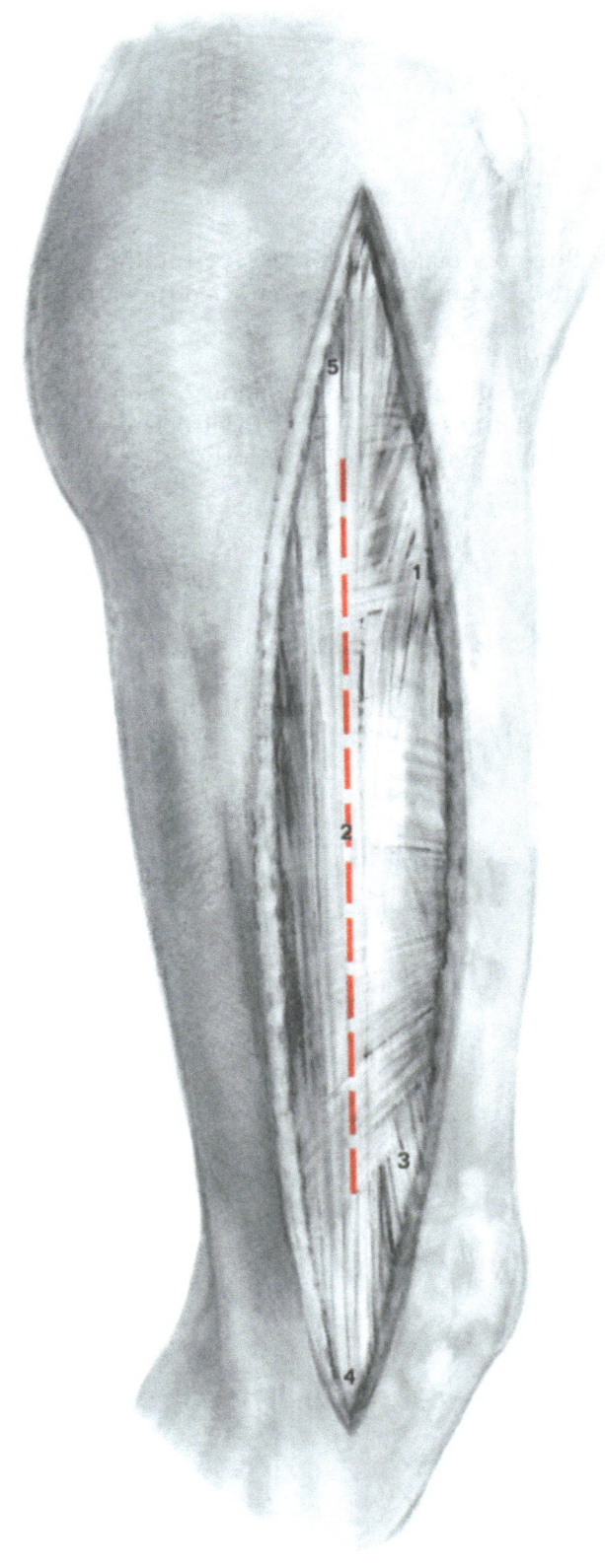

Fig. 18c

1   Tensor fasciae latae
2   Ilio-tibial tract
3   "Prepatellar tract" (of ilio-tibial tract)
4   Tubercle of GERDY
5   Greater trochanter

Deep Dissection   The three or four posterior or "perforating" vessels are divided between
(continued)   ligatures. The muscle is then retracted forward off the femoral shaft.

Extension   *Proximally*
The posterior half of the tendinous origin of the vastus lateralis muscle
is notched transversely below the trochanter, or it is divided completely
and the muscle is reflected anteriorly.

Fig. 18d

1   Greater trochanter
2   Gluteus medius muscle
3   Lateral circumflex femoral vessels
4   Frohse band
5   Rectus femoris muscle
6   Ilio-tibial tract
7   Vastus lateralis muscle

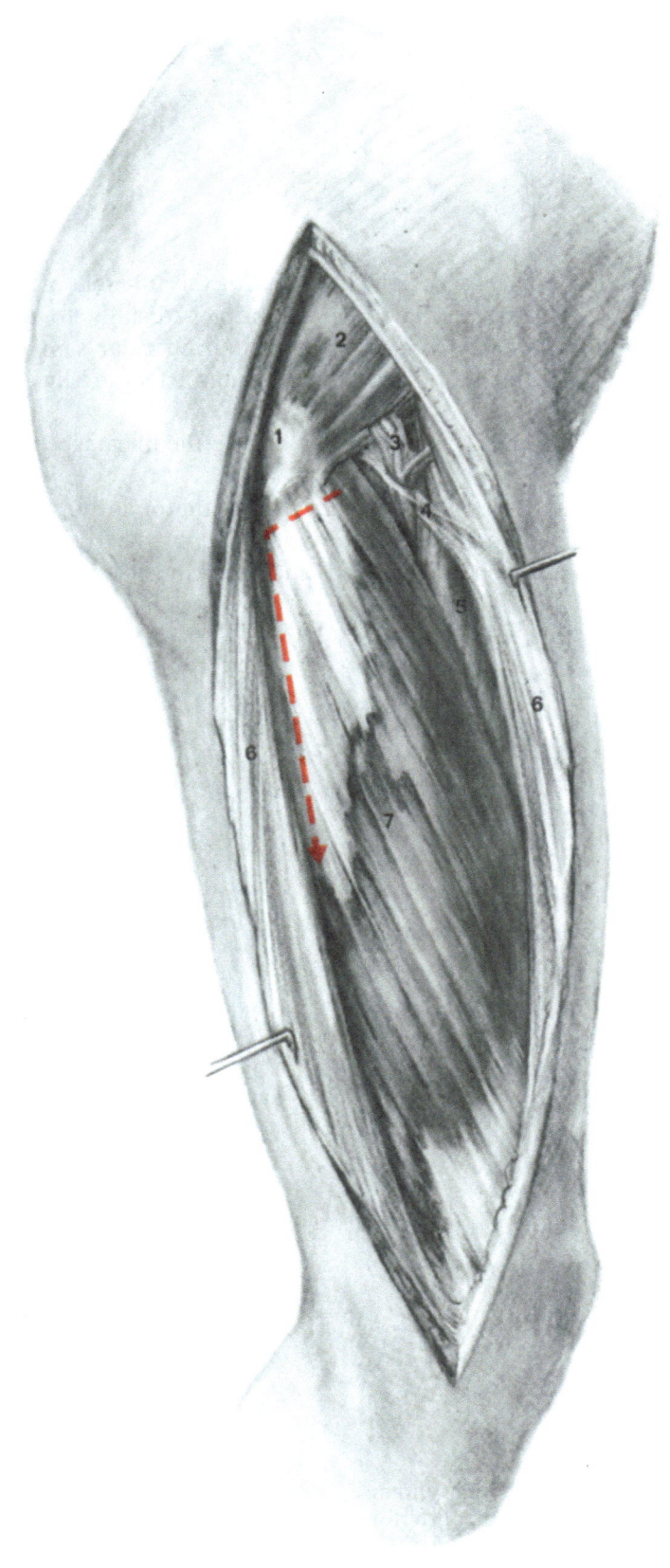

Extension   *Distally*
(continued) The ilio-tibial tract is split in line with the tubercle of GERDY. The muscle fibers of the vastus lateralis that attach distally to the aponeurosis are sharply separated from the lateral intermuscular septum. The perforating branches of the lateral superior genicular artery are ligated (see Chap. 19).

Internal Fixation   A plate is applied posterolaterally along the linea aspera.

Closure   The vastus lateralis is returned to its anatomic position and if necessary is reattached to its proximal origin below the trochanter. The ilio-tibial tract is readapted and sutured, and the skin is closed.

Hazards   None.

Fig. 18e

1 Vastus lateralis muscle
2 Lateral intermuscular septum
3 Perforating arteries, posterior vastal branches
4 Lateral superior genicular artery
5 Joint capsule
6 Lateral circumflex femoral vessels

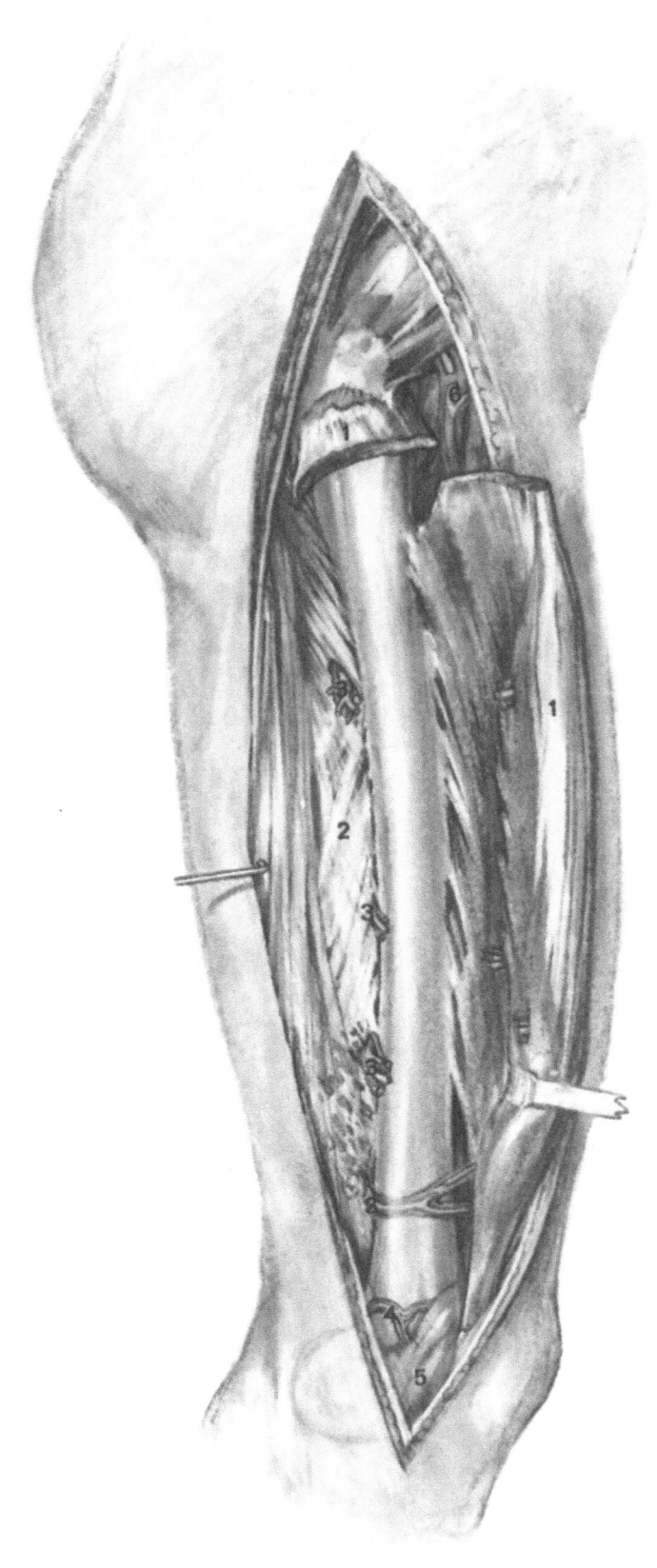

19 Distal Femur: Lateral Approach

Indications — Supracondylar or transcondylar fractures of the femur.

Positioning/ Draping — Supine with the knee flexed over a roll (60°–90°) to relax the tension of the gastrocnemius muscle. The hip and knee are draped free. Use of a sterile tourniquet may be indicated.

Skin Incision — The skin is incised along a line joining the *greater trochanter, lateral femoral condyle, and the tibial tuberosity*.

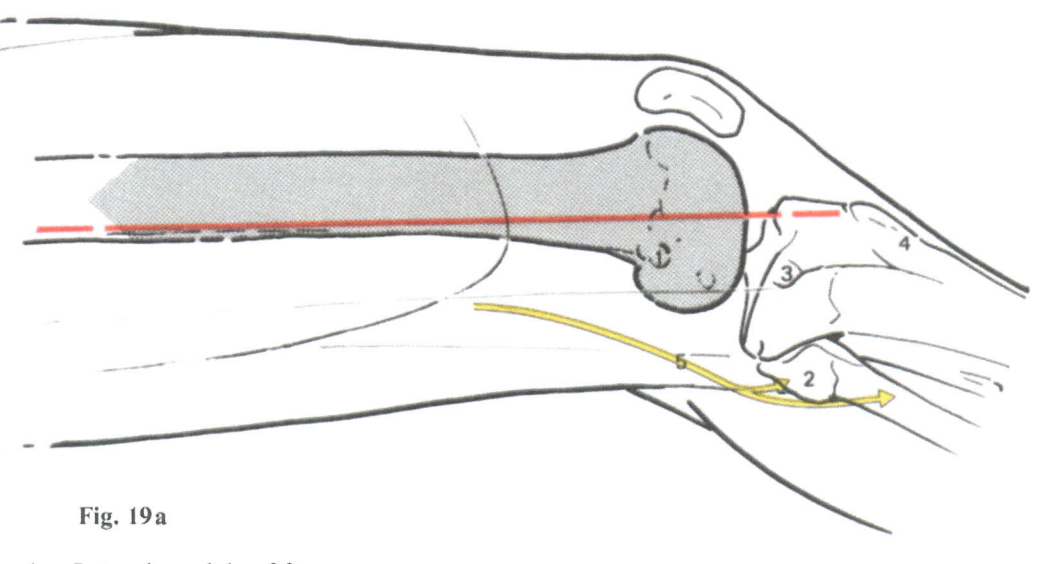

Fig. 19a

1 Lateral condyle of femur
2 Head of fibula
3 Tubercle of GERDY
4 Tuberosity of tibia
5 Common peroneal nerve

Deep Dissection   The ilio-tibial tract has its main distal attachment on the tubercle of GERDY of the tibia. Proximal to the lateral femoral condyle the tract divides in a V-shape, sending fibers to the patellar retinaculum ("prepatellar tract"). Starting at this bifurcation, the tract is split in a proximal direction, and the vastus lateralis is detached from the intermuscular septum and from the femur (see Chap. 18).

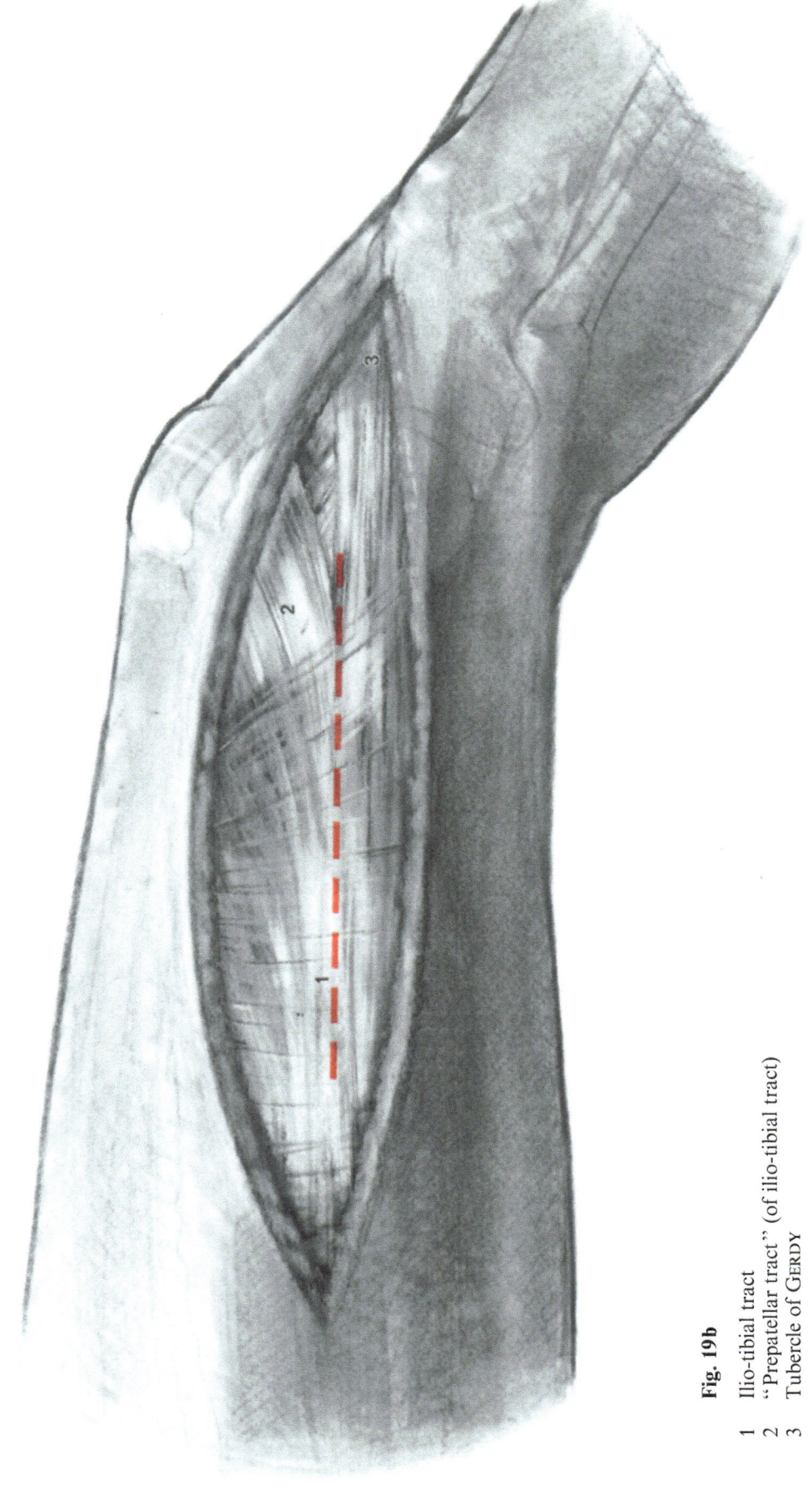

Fig. 19b

1 Ilio-tibial tract
2 "Prepatellar tract" (of ilio-tibial tract)
3 Tubercle of GERDY

Deep Dissection (continued)   The posterior or "perforating" vessels of the vastus lateralis and the proximal lateral genicular artery are ligated. To inspect the femoral condyle, the capsule is opened anterior to the palpable attachment of the lateral collateral ligament on the condyle.

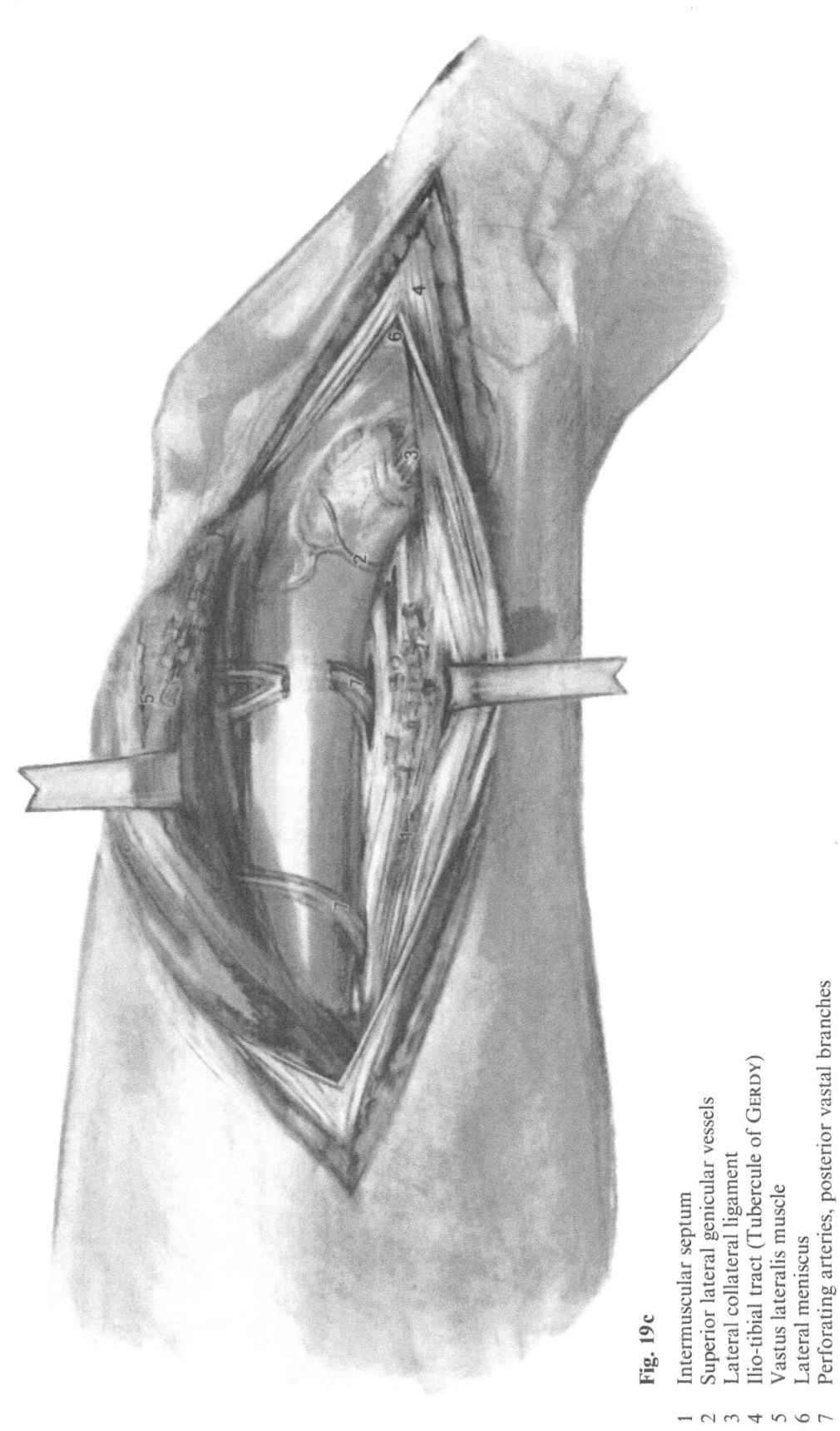

Fig. 19c

1 Intermuscular septum
2 Superior lateral genicular vessels
3 Lateral collateral ligament
4 Ilio-tibial tract (Tubercule of GERDY)
5 Vastus lateralis muscle
6 Lateral meniscus
7 Perforating arteries, posterior vastal branches

135

19 Distal Femur: Lateral Approach

Extension   *Distally and medially*
The skin incision is extended distally to 2 fingerwidths below the tibial tuberosity, keeping lateral to the tibial crest. The tibial tuberosity is osteotomized, or the patellar ligament is divided in a Z-shaped fashion. The patella is reflected medially and upward to expose both femoral condyles (this may require detachment of the infrapatellar fatty body).

Proximally
The incision is carried upward toward the greater trochanter (see Chap. 18).

Internal Fixation   With blade or screws passing through the condyle, the side plate is applied to the postero-lateral aspect of the femur.

Closure   The joint capsule is closed by suture, and the ilio-tibial tract is readapted. The osteotomized tibial tuberosity is reattached with a cancellous bone screw, or the cut patellar ligament is sutured, and the skin is closed.

Fig. 19d

1   Tuberosity of tibia (after osteotomy)
2   Patella, articular surface
3   Suprapatellar recess
4   Articular capsule (cut)
5   Infrapatellar fatty body of HOFFA

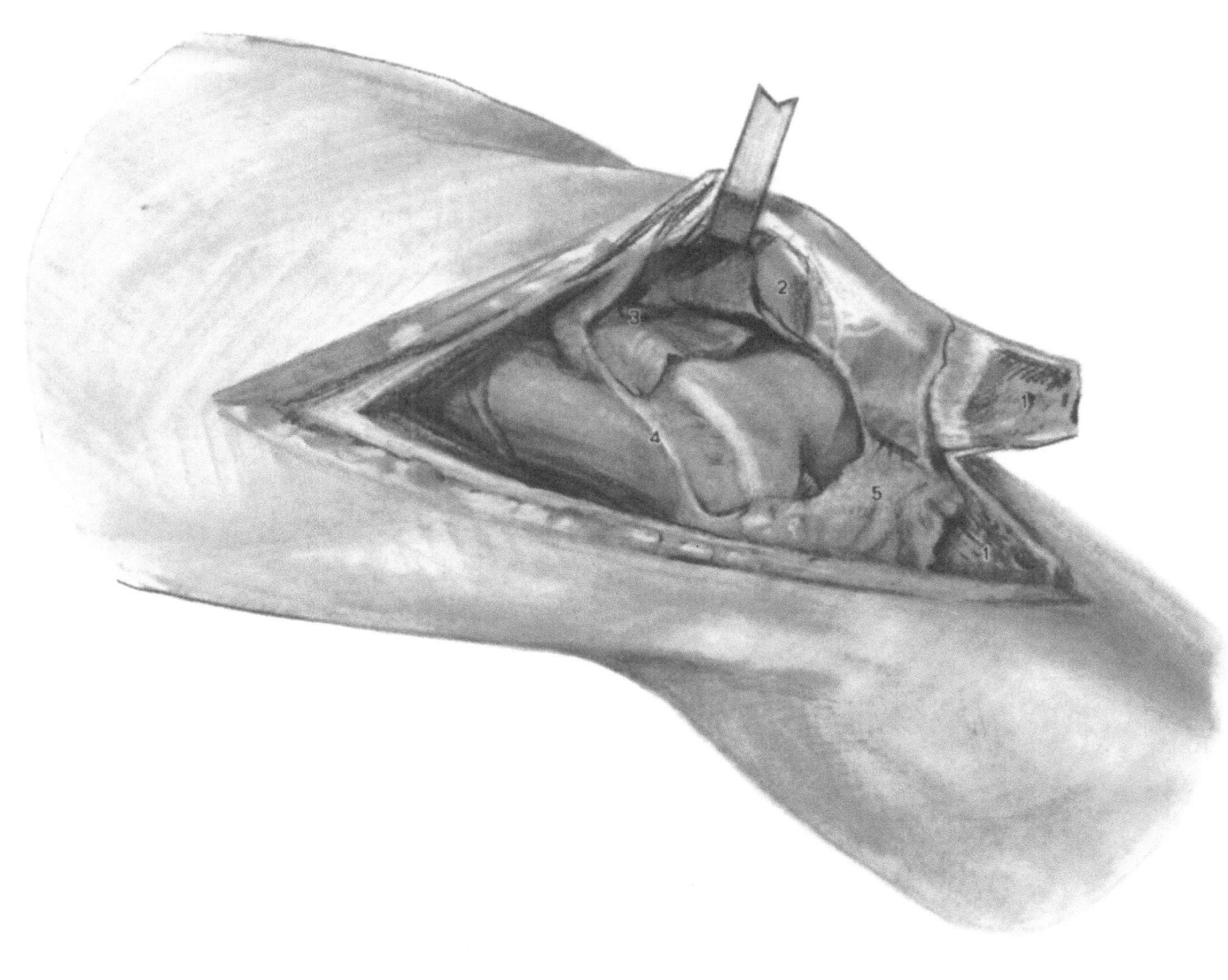

20 Patella

Indications   Fractures of the patella.

Positioning   Supine with the leg extended; a tourniquet is applied.

Skin Incision   A transverse incision is made across the midportion of the patella, extending to the condyles on each side. Alternatively, a lateral parapatellar incision may be used (see Chap. 21).
In the presence of an open fracture or chronic inflammation, the subcutaneous prepatellar bursa is excised.

For medullary nailing of the tibia
A transverse or vertical incision halfway between the lower edge of the patella and the tibial tuberosity is made. The patellar ligament is split longitudinally in order to expose the point of entry of the nail.

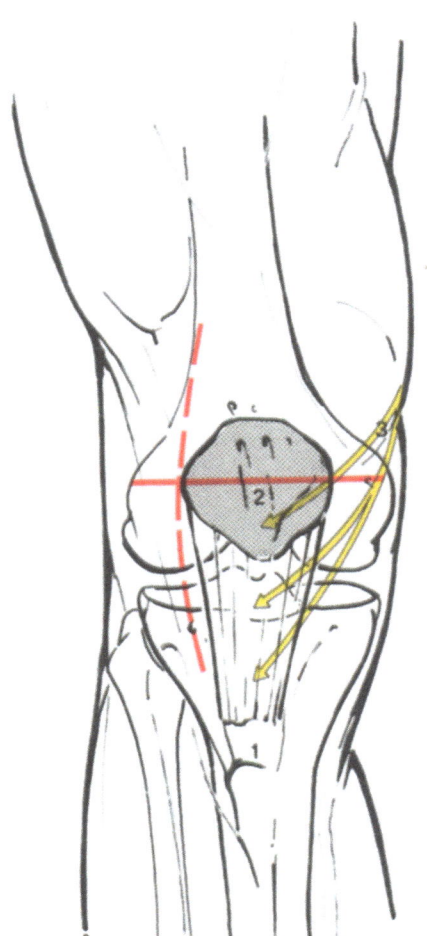

Fig. 20a

1   Tuberosity of tibia
2   Patellar ligament
3   Infrapatellar branches
    of saphenous nerve

Deep Dissection   It is seldom necessary since the fracture is superficial and only the superior and inferior poles of the patella have to be identified.

Extension   If a small distal fragment has been avulsed, it is advisable to secure the patellar ligament to the tibial tuberosity with a wire loop. For this, a transverse skin incision about 2–3 cm long is made over the tuberosity, and the wire is drawn into place by a subcutaneous pull-through technique.

Closure   The parapatellar retinacula are sutured and the skin is closed.

Fig. 20 b

1   Subcutaneous prepatellar bursa
2   Infrapatellar branches of saphenous nerve

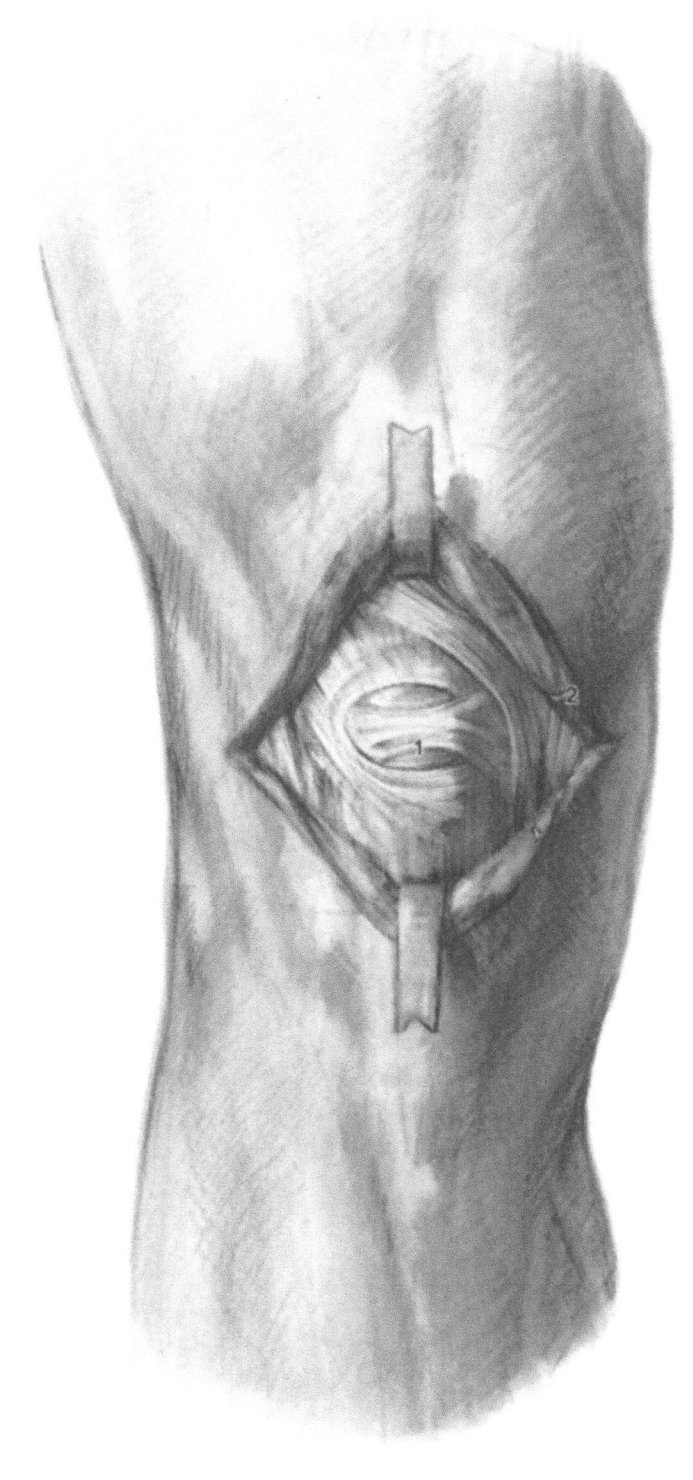

21 Knee Joint and Proximal Tibia: Lateral Parapatellar Approach

Indications Fractures of the tibial plateau, simple fractures of the lateral femoral condyle, recent or old injuries of the cruciate or collateral ligaments.

Advantages The lateral parapatellar approach gives a good exposure for both lateral and medial injuries. It is readily extensible, and causes fewer healing problems than angled incisions. Moreover, it poses no danger to the major nerve supply to the skin of the knee arising from the medially situated saphenous nerve. The same incision may be used for arthroplasty of the knee (as a secondary procedure) (see W. MÜLLER 1982).

Positioning/ Draping Supine with the knee flexed 50°–60° over a roll. The leg and knee are draped so that they can be freely moved, tourniquet.

Skin Incision With the leg extended, it starts over the distal head of the *vastus lateralis,* is carried straight downward just lateral to the *patella,* and terminates 2–3 cm distal and slightly lateral to the *tibial tuberosity.*

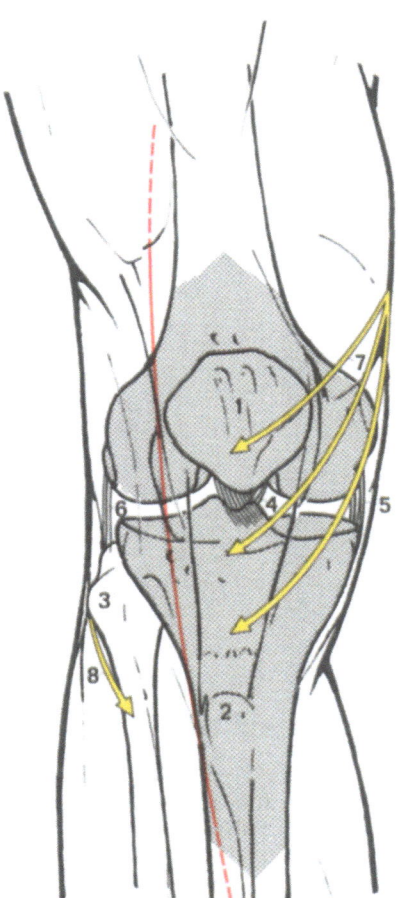

Fig. 21a

1 Patella
2 Tuberosity of tibia
3 Head of fibula
4 Cruciate ligaments
5 Medial collateral ligament
6 Lateral collateral ligament
7 Infrapatellar branches
  of saphenous nerve
8 Superficial peroneal nerve

Deep Dissection   *For fractures of the lateral tibial plateau*, the aponeurotic fascia covering the extensor muscles is incised 1 cm lateral to the anterior tibial crest. The incision is carried proximally through the longitudinal patellar retinaculum to the anterior margin of the ilio-tibial tract.

Fig. 21b

1   "Prepatellar tract" of ilio-tibial tract
2   Ilio-tibial tract
3   Tubercle of GERDY
4   Tuberosity of tibia
5   Quadriceps tendon

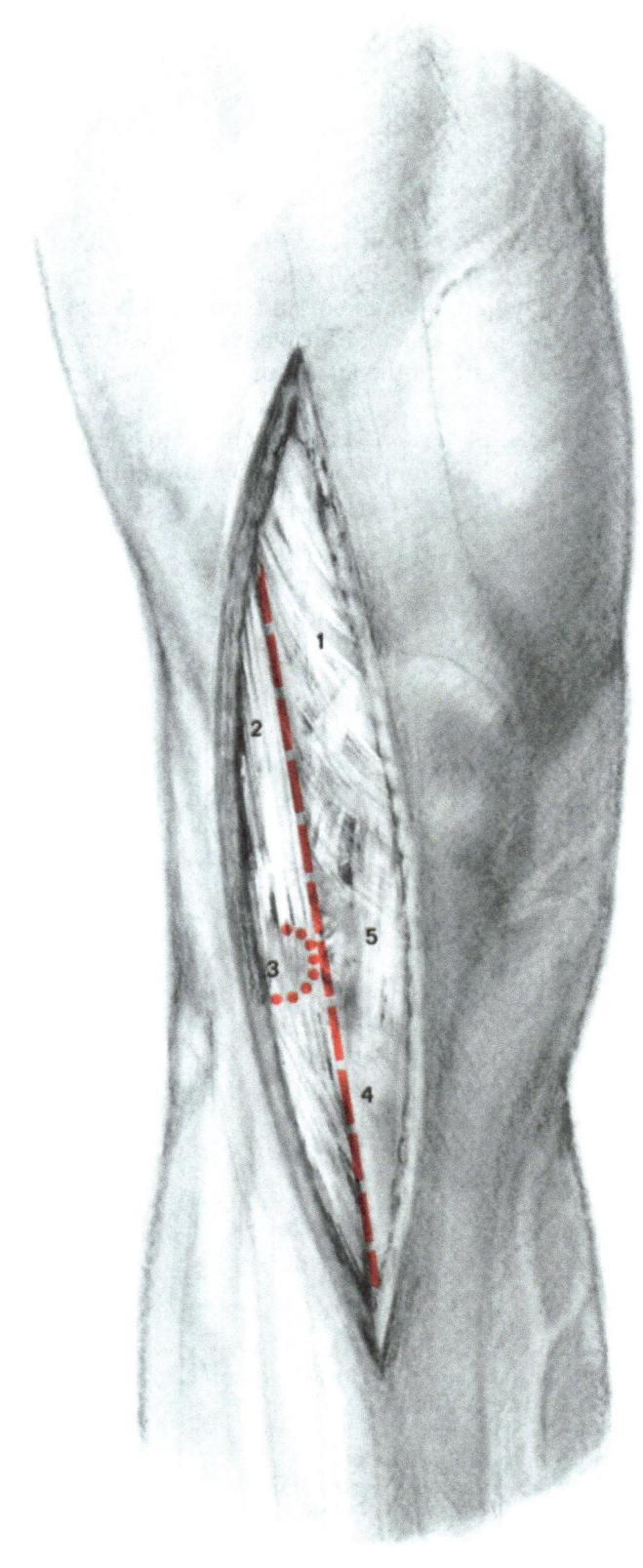

Deep Dissection
(continued)

The tubercle of GERDY is superficially osteotomized, and the ilio-tibial tract as well as the entire tibial extensor muscle group are detached back to the lateral collateral ligament and retracted laterally.

The lateral tibial articular surface is exposed by opening the joint capsule transversely between the lateral meniscus and tibial condyle.

Internal Fixation

A T-plate, L-plate, or DCP is applied to the proximal tibia beneath the extensor muscle group to buttress the lateral tibial plateau.

Fig. 21c

1   Ilio-tibial tract
2   Infrapatellar fatty body of HOFFA
3   Lateral condyle of tibia
4   Tubercle of GERDY (after osteotomy)
5   Anterior tibial recurrent artery
6   Muscular branch of deep peroneal nerve
7   Extensor muscles (anterior tibialis muscle)
8   Tuberosity of tibia
9   Infrapatellar branches of saphenous nerve
10 Crural fascia
11 Lateral meniscus (elevated)

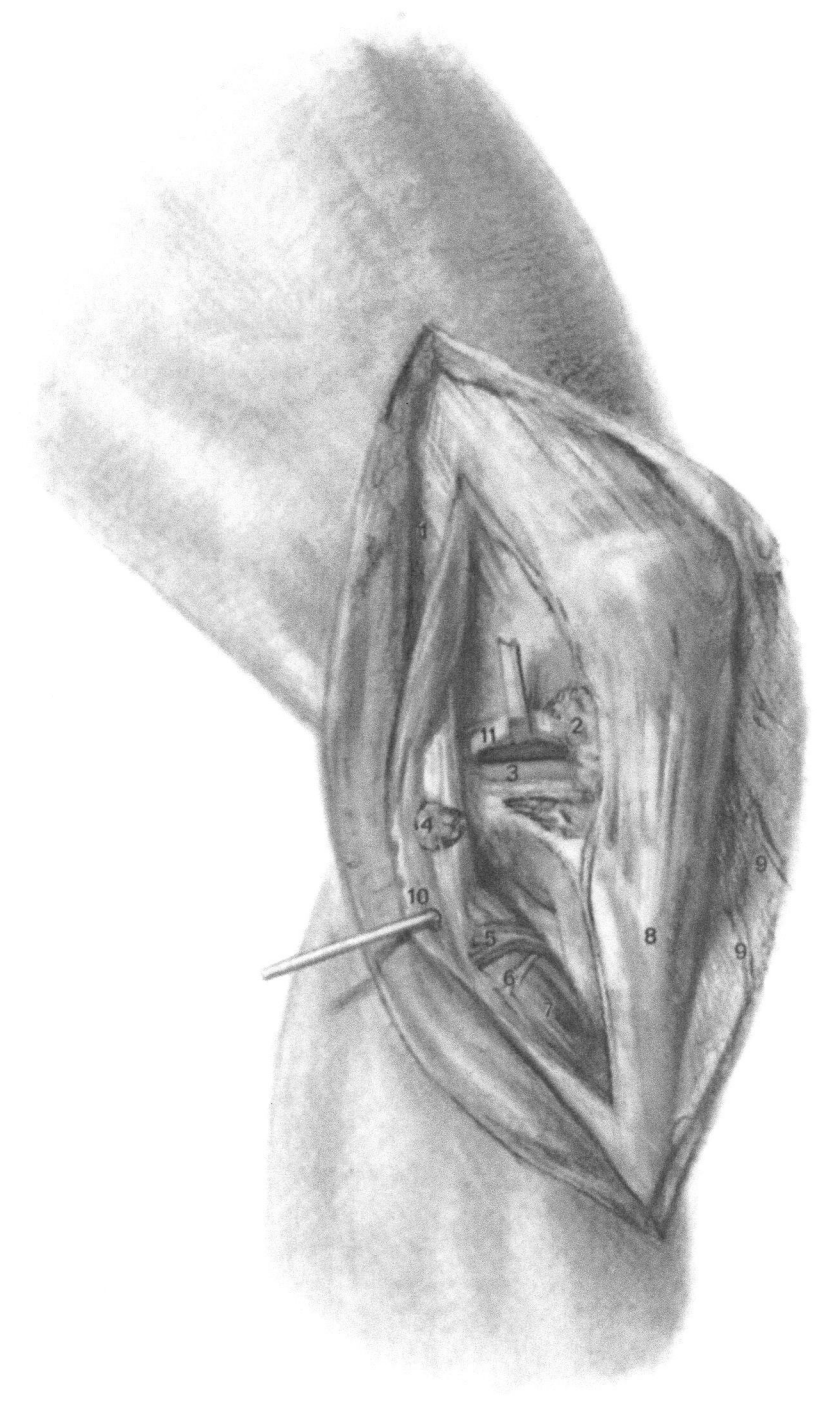

Extension   *To the medial condyle.*
The skin flap including the superficial fascia and the subcutaneous bursa is retracted medially past the anterior surface of the patella. Osteotomy of the tibial tuberosity or Z-shaped division of the patellar ligament is seldom necessary.

Fig. 21d

1   Ilio-tibial tract
2   Prepatellar bursa
3   Infrapatellar branches of saphenous nerve
4   Saphenous nerve
5   Great saphenous vein
6   Pes anserinus
7   Medial collateral ligament

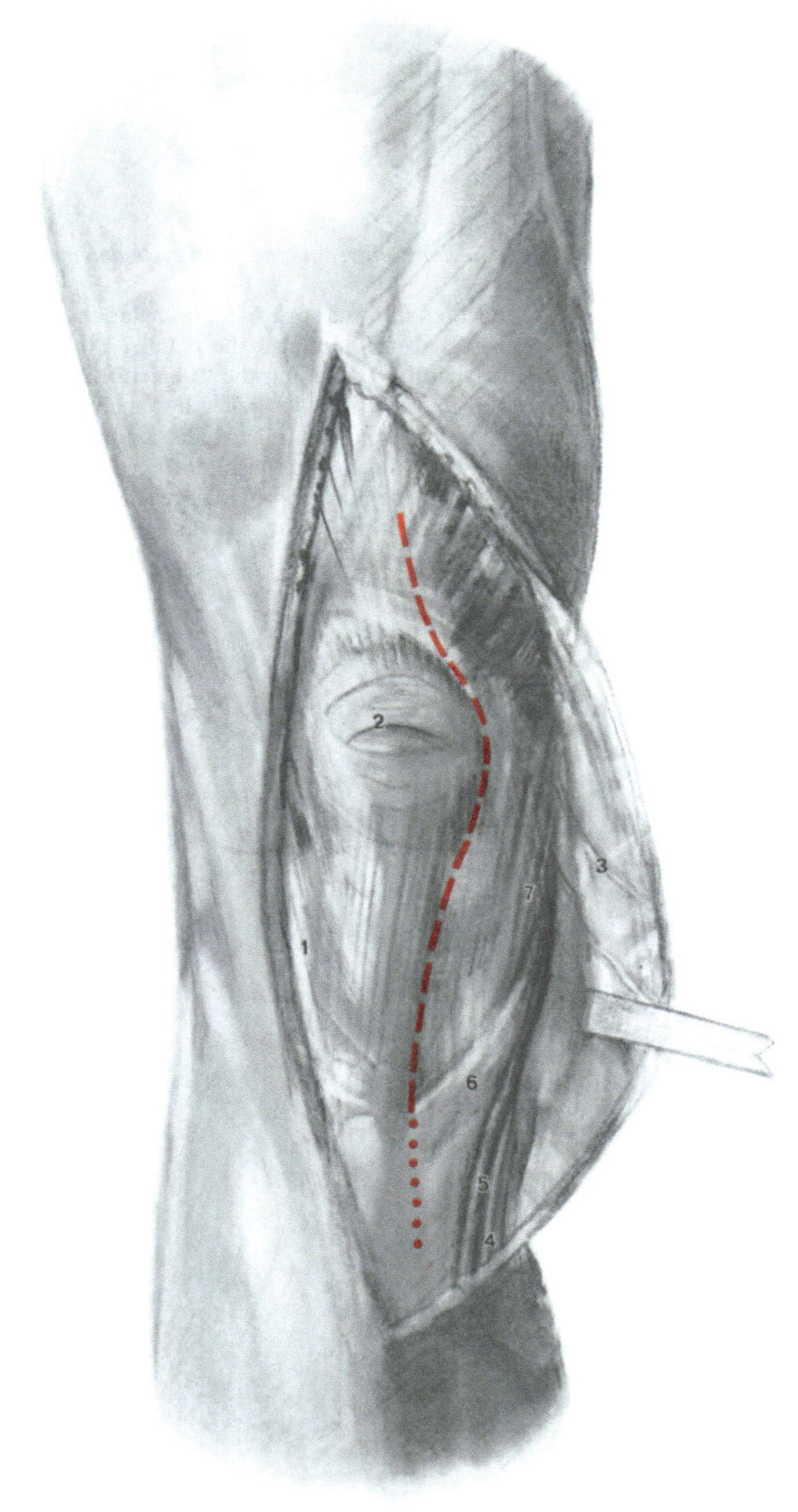

Deep Dissection   *For a medial arthrotomy to repair the cruciate ligaments or medial collateral ligament.*

The skin flap together with the superficial fascia and subcutaneous prepatellar bursa are dissected medially past the patella. By flexing the knee, even postero-medial structures (including the "posterior corner" of the capsule) can be made accessible. The joint is opened by a medial, longitudinal parapatellar incision. Notching the vastus medialis tendon enables the patella to be dislocated laterally. In repair of the anterior cruciate ligament the posterior aspect of the lateral femoral condyle is reached by blunt dissection between the ilio-tibial tract and vastus lateralis.

Fig. 21e

1 Quadriceps tendon
2 Quadriceps femoris, vastus medialis
3 Suprapatellar recess
4 Infrapatellar fatty body of Hoffa
5 Tuberosity of tibia
6 Anterior cruciate ligament
7 Posterior cruciate ligament
8 Anterior horn of medial meniscus
9 Pes anserinus

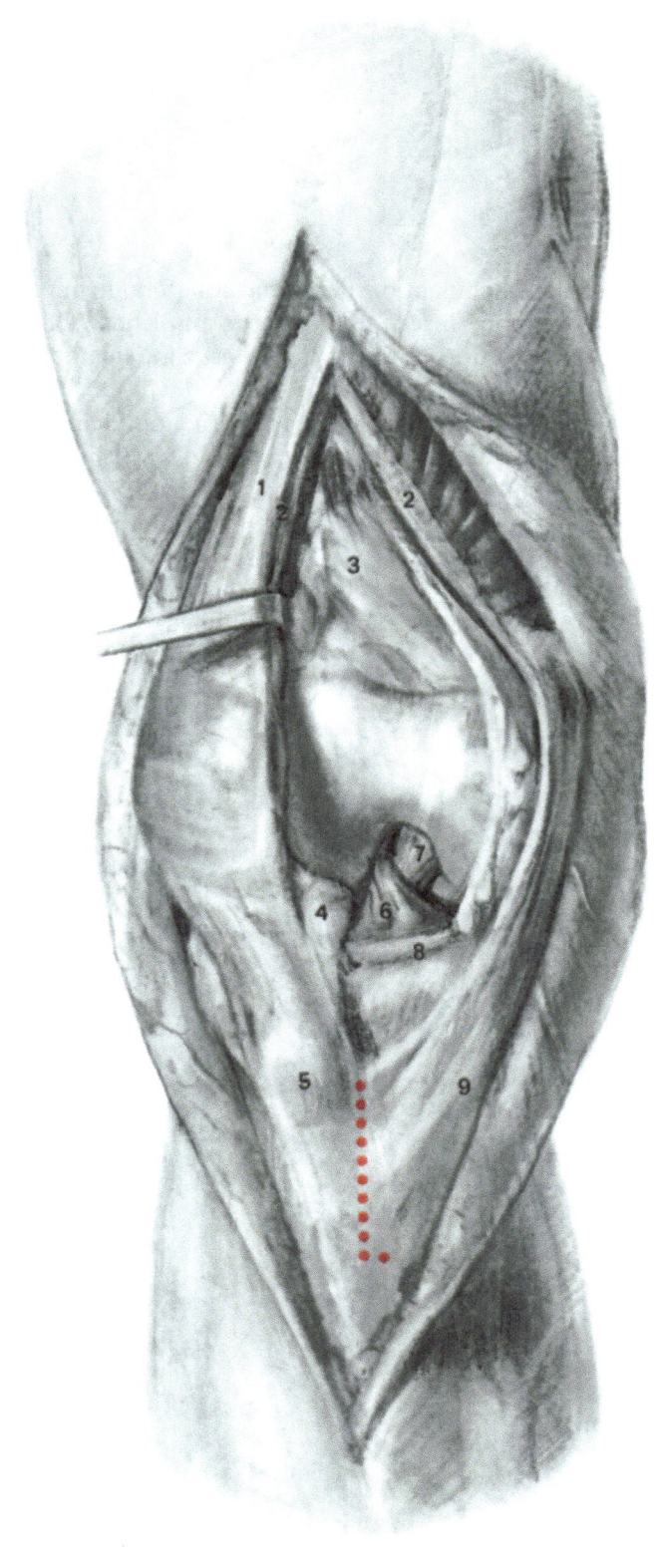

Deep Dissection   The tibial attachment of the medial collateral ligament is reached by osteoto-
(continued)   mizing the bony insertion of the pes anserinus. This also enables to palpate
the tibial insertion of the posterior cruciate ligament.

Closure   Detached menisci are resutured and the joint capsule is closed. The tubercle
of GERDY is reattached with a screw and washer. Incisions in the aponeurosis
and ilio-tibial tract are sutured.
The bony attachment of the pes anserinus is readapted with a screw and
washer.
The wound is closed only after the tourniquet has been released and a
meticulous hemostasis has been obtained.

Hazards   None.

Fig. 21f

1   Quadriceps femoris, vastus medialis
2   Suprapatellar recess
3   Infrapatellar fatty body of HOFFA
4   Anterior cruciate ligament
5   Medial proximal genicular artery
6   Medial collateral ligament
7   Pes anserinus (after osteotomy)
8   Medial meniscus and joint capsule
9   Medial distal genicular artery

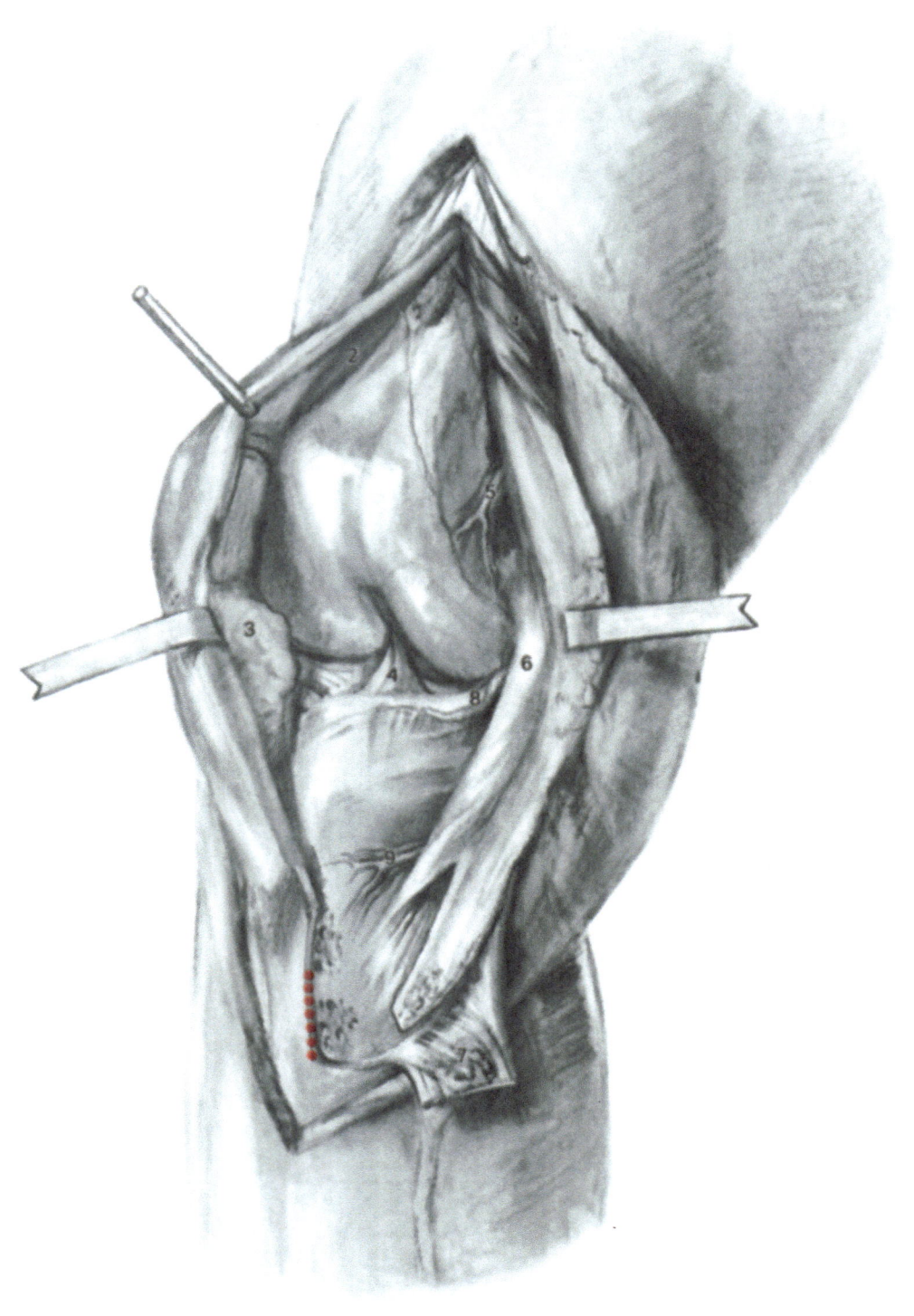

22 Tibial Shaft: Anterior Approach

Indications	Fractures or non-unions of the tibial shaft.
Positioning/Draping	Supine with the tibia supported and in slight internal rotation; tourniquet.
Skin Incision	Parallel and 10–15 mm lateral to the anterior tibial crest, extending from the tibial tuberosity to 3 fingerwidths above the ankle joint.

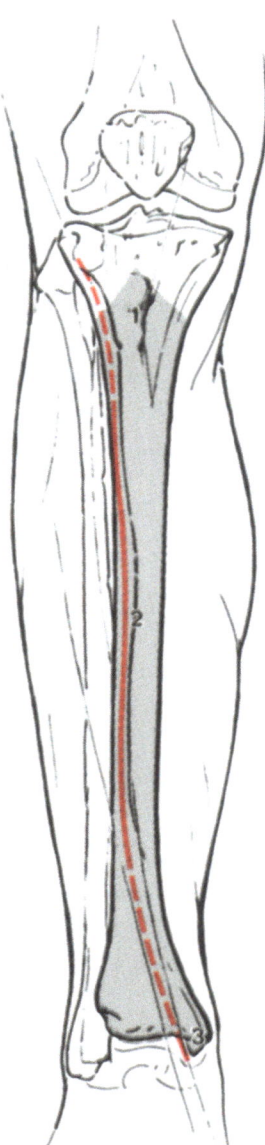

Fig. 22a

1 Tuberosity of tibia
2 Anterior border
3 Medial malleolus

Deep Dissection   *For medial plating (closed fracture)*
The crural fascia or tendon compartment of the tibialis anterior muscle, if not already torn, is opened so that reduction can be carried out (temporary cerclage or bone-holding clamps). The plate should be applied as close as possible to the posterior edge of the medial tibial surface, as the soft-tissue covering is better there than anteriorly.

For lateral plating (open fracture or skin contusion)
The crural fascia is opened longitudinally 10–15 mm lateral to the anterior tibial crest. The tibialis anterior muscle is retracted from the bone when the plate is applied.

Extension   *Proximally*
Laterally past the tibial tuberosity to the outer margin of the patella (see Chap. 21).

Distally
Curves gently to the tip of the medial malleolus (see Chap. 24). The great saphenous vein may have to be divided and ligated.

Closure   The crural fascia is approximated to the anterior border of the tibia if possible, and the skin is closed.

Hazards   None.

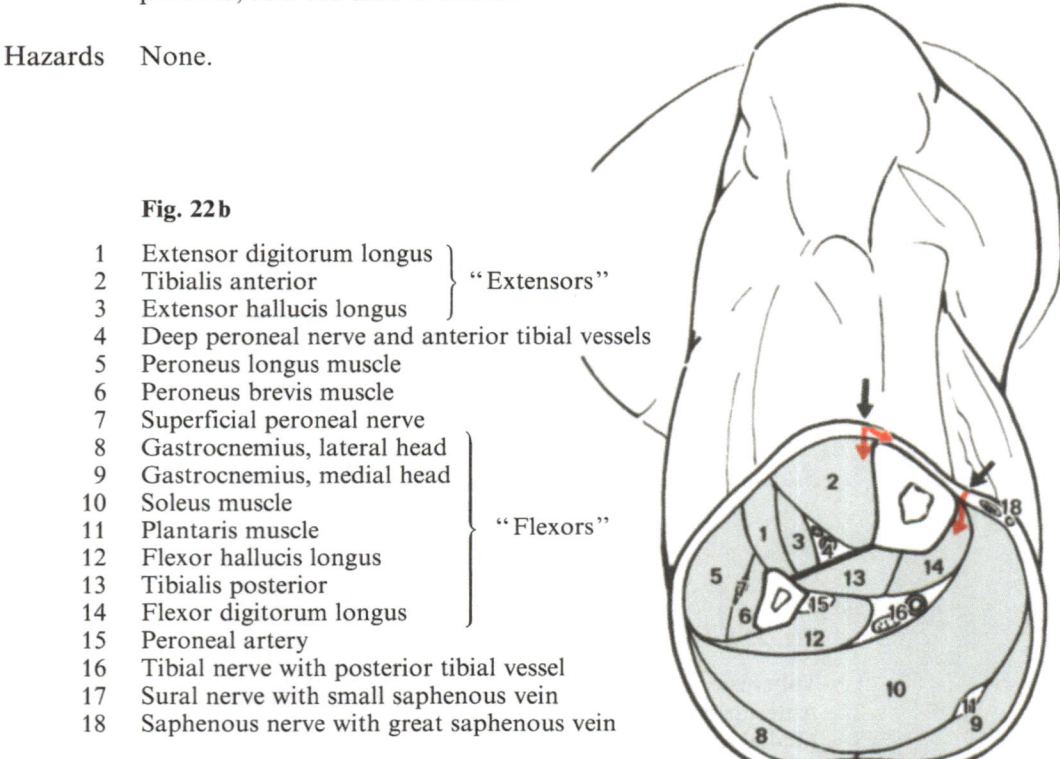

Fig. 22 b

1   Extensor digitorum longus ⎫
2   Tibialis anterior   ⎬ "Extensors"
3   Extensor hallucis longus ⎭
4   Deep peroneal nerve and anterior tibial vessels
5   Peroneus longus muscle
6   Peroneus brevis muscle
7   Superficial peroneal nerve
8   Gastrocnemius, lateral head ⎫
9   Gastrocnemius, medial head ⎪
10 Soleus muscle   ⎪
11 Plantaris muscle   ⎬ "Flexors"
12 Flexor hallucis longus   ⎪
13 Tibialis posterior   ⎪
14 Flexor digitorum longus   ⎭
15 Peroneal artery
16 Tibial nerve with posterior tibial vessel
17 Sural nerve with small saphenous vein
18 Saphenous nerve with great saphenous vein

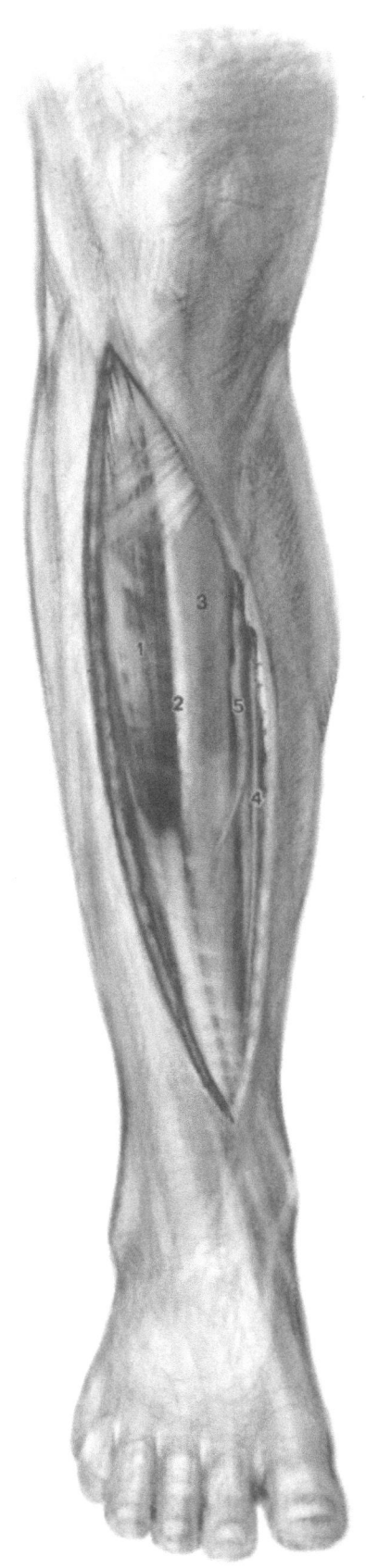

Fig. 22c

1   Tibialis anterior muscle
2   Anterior crest
3   Medial surface of tibia
4   Saphenous nerve
5   Great saphenous vein

23 Tibial Shaft: Postero-medial Approach

Indications Shaft fractures, non-unions, or infections associated with poor soft-tissue conditions on the anterior side of the lower leg; particularly indicated for secondary procedures, and compartement syndroms.

Positioning Supine with the hip in extreme external rotation, the knee flexed, and the foot resting on the contralateral tibia.

Skin Incision Extends from *behind the upper tibia* to behind the medial malleolus, parallel and 1–2 fingerwidths posterior to the posterior tibial border (*beware* of injury to great saphenous vein and saphenous nerve).

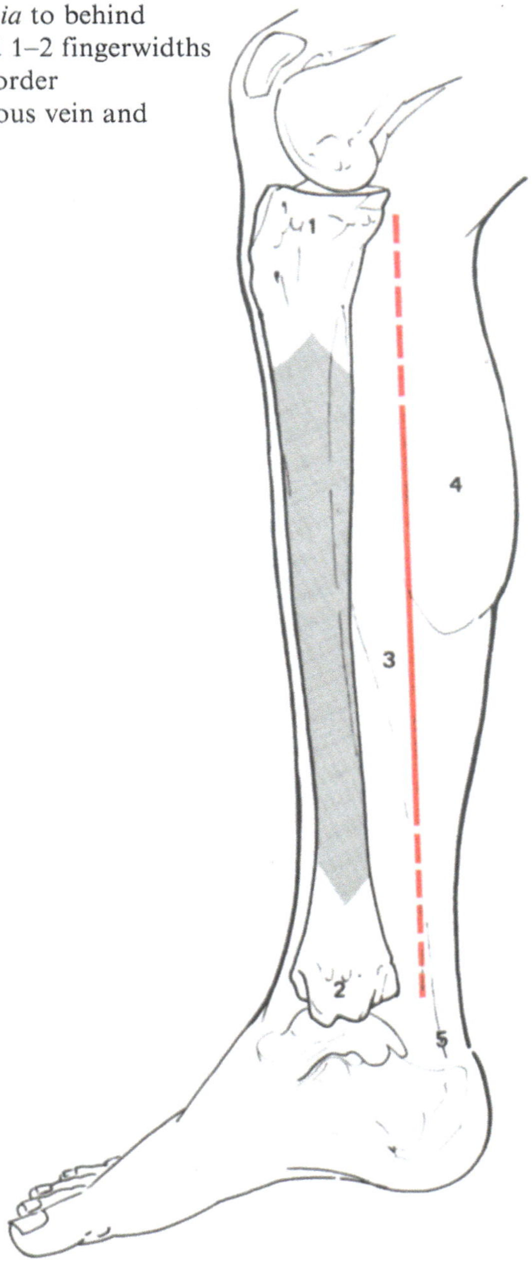

Fig. 23a

1 Medial condyle of tibia
2 Medial malleolus
3 Soleus muscle
4 Gastrocnemius muscle
5 Calcaneal tendon (Achilles tendon)

Deep Dissection   The *superficial* flexor compartment is opened by splitting the superficial crural fascia parallel to the posterior tibial crest (sparing the great saphenous vein and saphenous nerve). The origin of the soleus muscle is identified on the tibia.

Fig. 23b

1   Saphenous nerve
2   Great saphenous vein
3   Crural fascia (superficial layer)
4   Medial condyle of tibia
5   Medial malleolus

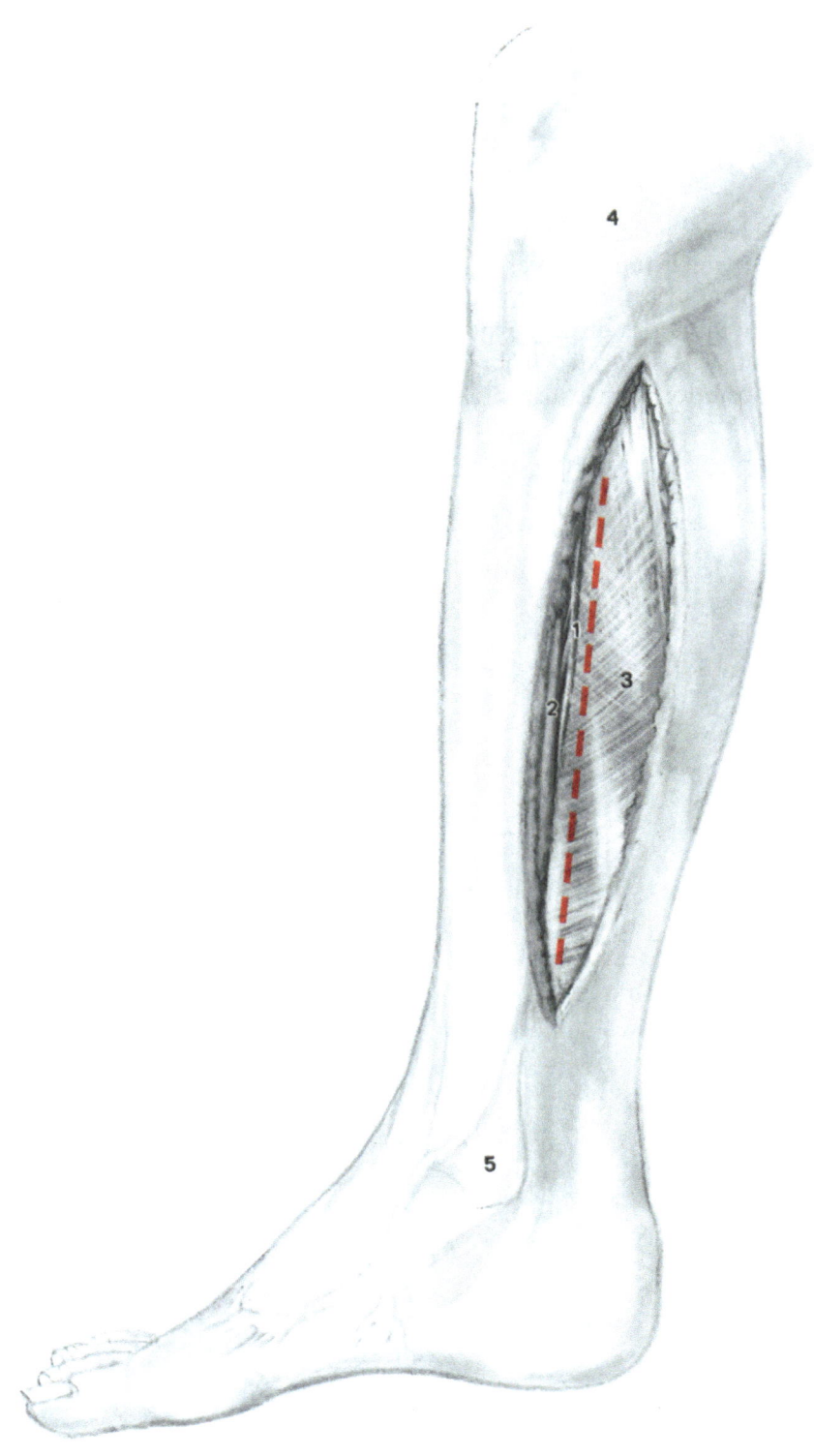

Deep Dissection
(continued)

The *deep* flexor compartment is opened by splitting the deep crural fascia, at which time the perforating veins are ligated. Distal to the origin of the soleus lies the flexor digitorum longus, covering the posterior surface of the tibia. Its distal portion can be retracted, while its proximal attachment must be separated from the tibia by sharp dissection. The adjacent tibialis posterior muscle also can be retracted postero-laterally.

Fig 23c

1   Crural fascia
    (superficial layer)
2   Crural fascia
    (deep layer/flexor digitorum longus)
3   Gastrocnemius muscle, medial head
4   Soleus muscle
5   Perforating veins

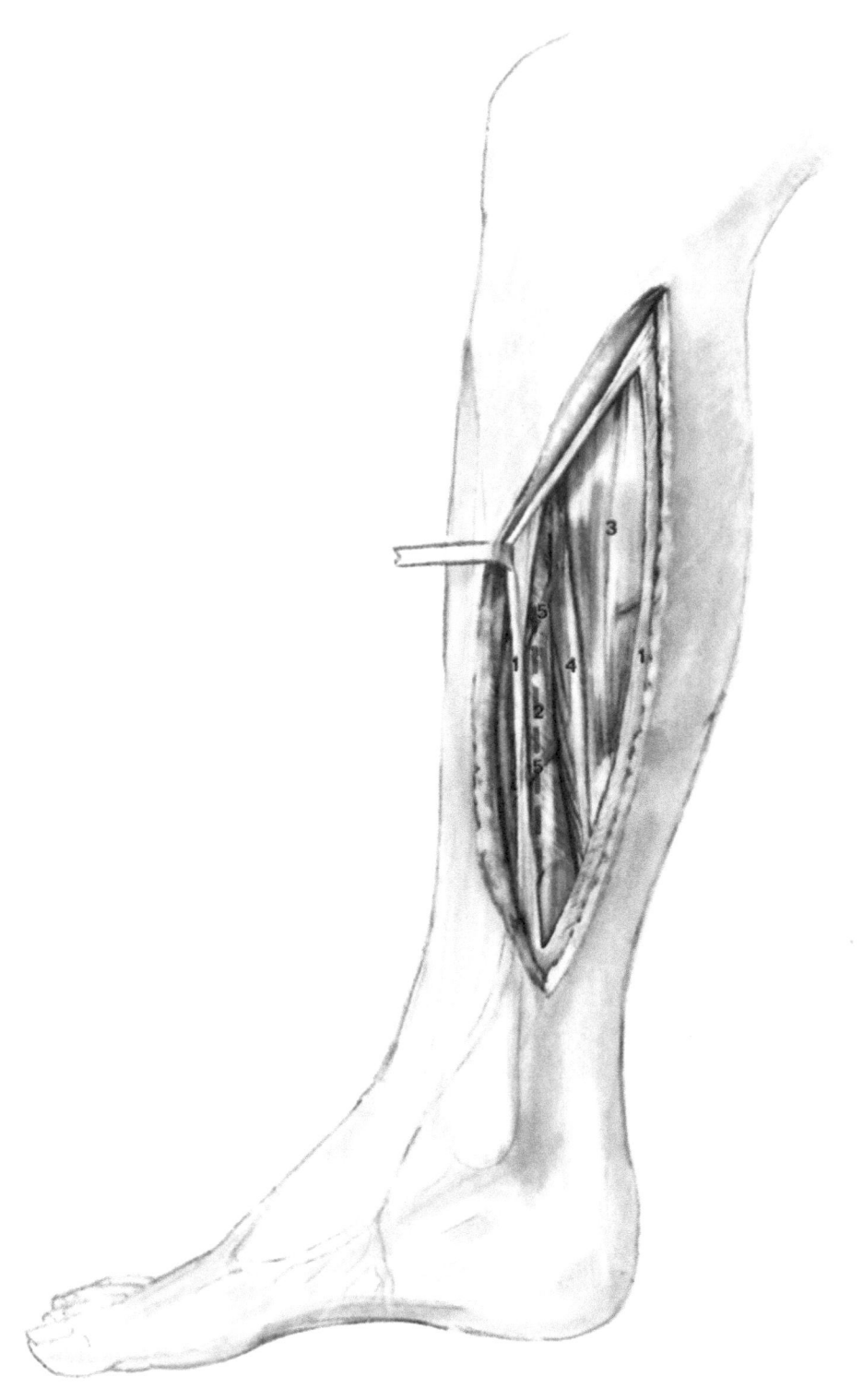

23   Tibial Shaft: Postero-medial Approach

Extension   Practically none.

Internal Fixation   A plate is applied to the postero-medial surface of the tibia.

Closure   It may be necessary to reapproximate the superficial layer of the crural fascia.

Hazards   Injury to the posterior neurovascular bundle.

Fig. 23d

1   Flexor digitorum longus muscle
2   Soleus muscle
3   Gastrocnemius muscle, medial head
4   Tibialis posterior, tendon ("chiasma crurale")
5   Crural fascia (superficial) layer)

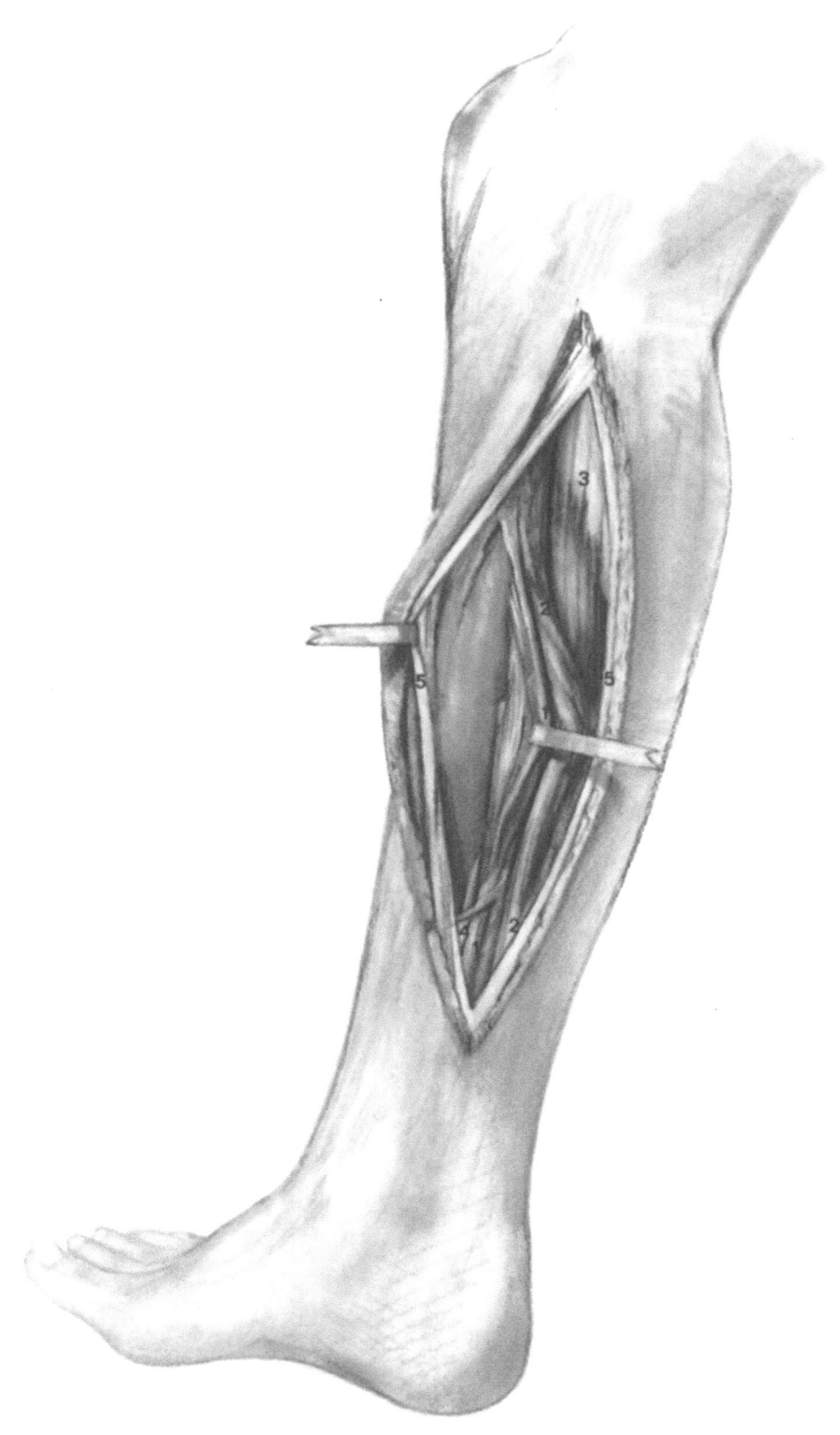

24 Medial Malleolus and Distal Tibia

Indications   Bony or ligamentous injuries of the medial malleolus and distal tibia (the "tibial pilon").

Positioning   Supine; tourniquet.

Skin incision   *Malleolar fractures*
Curves around and 1–2 fingerwidths from the anterior side of the medial malleolus to a point 1–2 cm distal to the tip of the malleolus.

Pilon fractures
The incision is extended proximally, passing lateral to the anterior crest of the tibia.

Note: When a second incision is made for internal fixation of the fibula, a bridge of skin at least 5 cm wide should be left between the medial and lateral incisions!

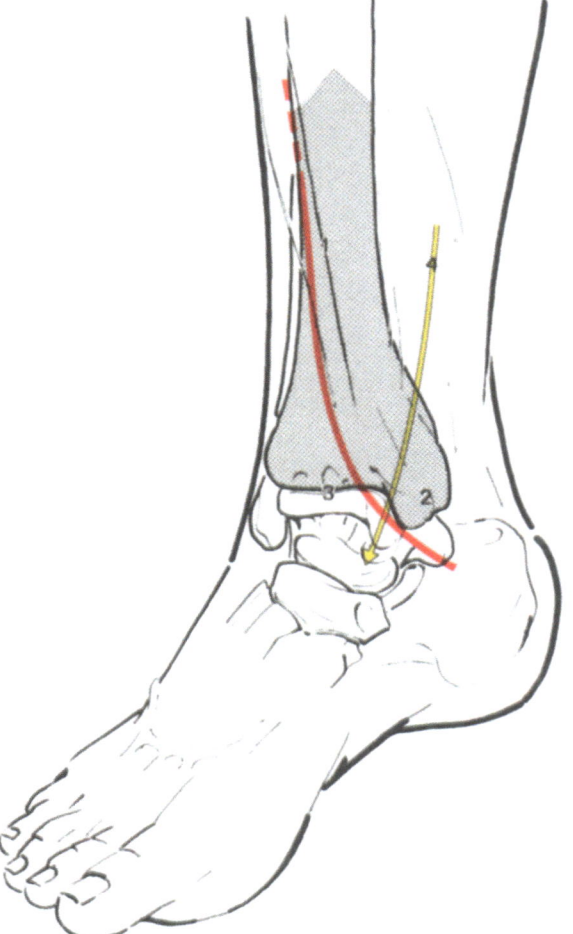

Fig. 24a

1   Anterior border of tibia
2   Medial malleolus
3   Ankle joint (talocrural joint)
4   Saphenous nerve

Deep Dissection   The great saphenous vein and saphenous nerve should be spared if at all possible. Most malleolar fractures will lie just below the skin.

Internal Fixation   *Malleolar fracture*
Ordinarily, one or two 3.5-mm screws are inserted through the tip of the medial malleolus. A tension band may be needed in some cases.

Fig. 24 b

1   Great saphenous vein
2   Saphenous nerve
3   Crural fascia
4   Trochlea of talus
5   Medial malleolus
6   Medial ligament (deltoid ligament)

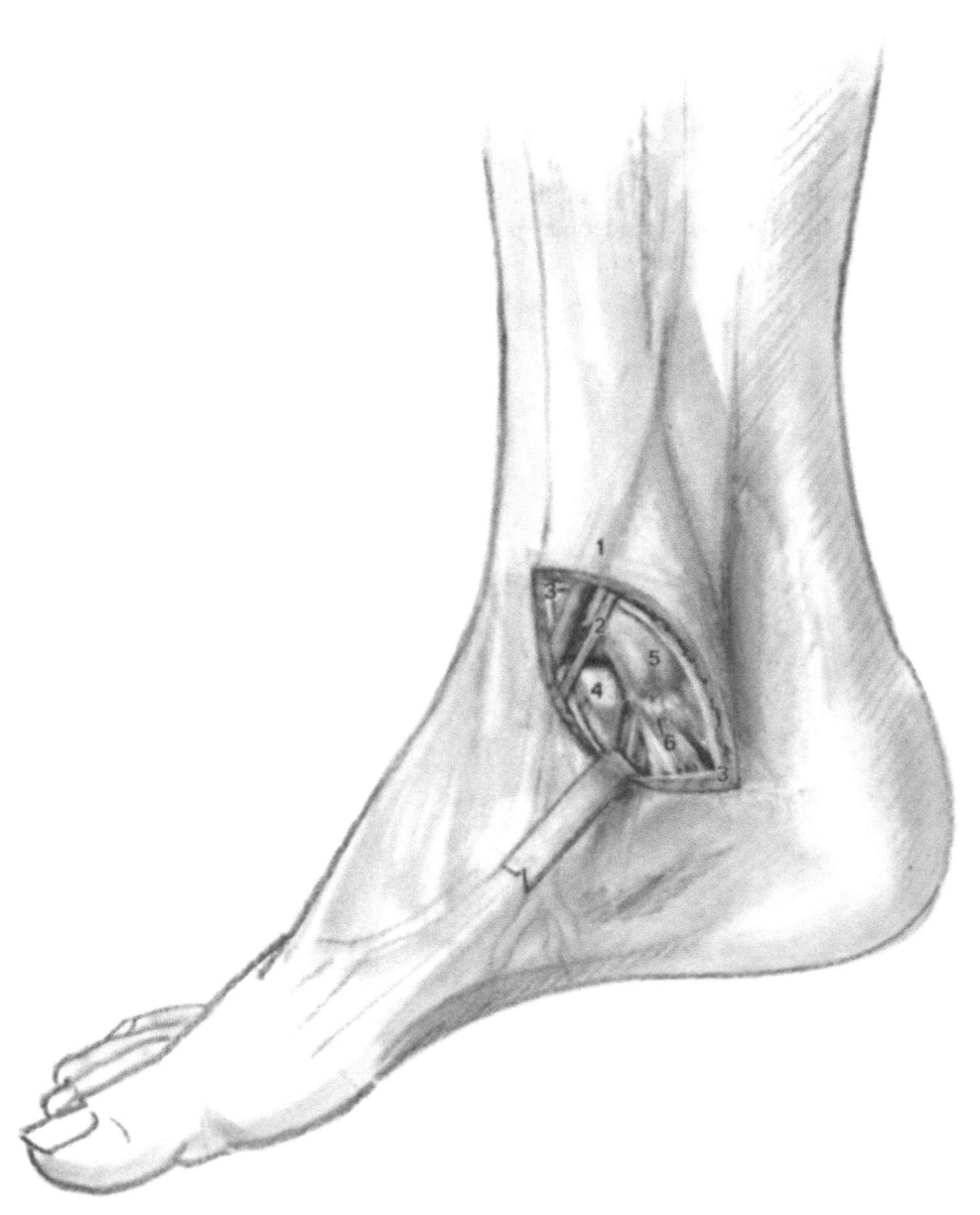

Deep Dissection
(continued)

In tibial pilon fractures, it is necessary to fenestrate the anterior joint capsule in order to inspect the ankle joint. By upward and lateral retraction of the tendon of the tibialis anterior and of the neurovascular bundle (anterior tibial vessels) and deep peroneal nerve) the lateral half of the distal tibia may be visualised.

Fig. 24c

1   Tibialis anterior, tendon
2   Saphenous nerve
3   Great saphenous vein
4   Medial malleolus
5   Superior extensor retinaculum

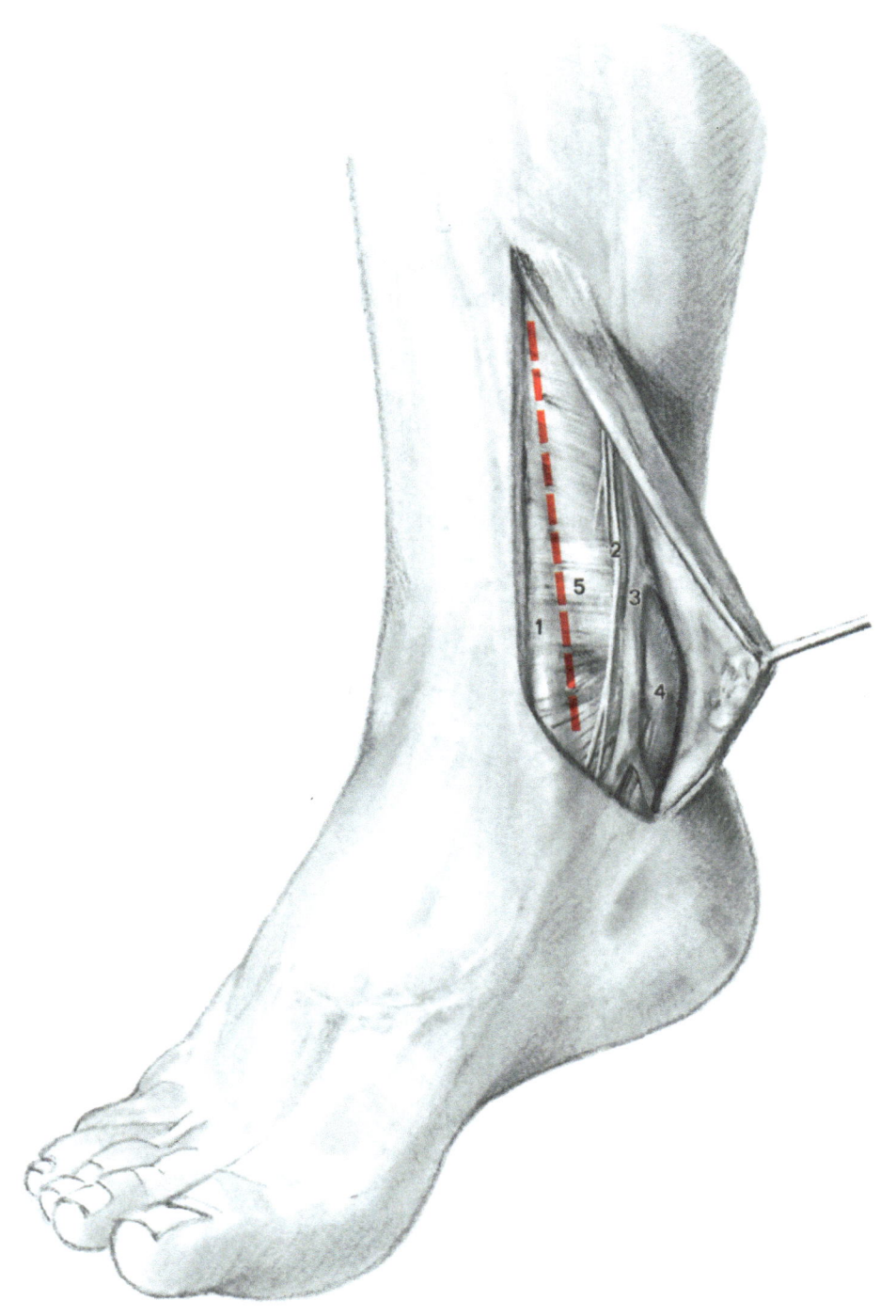

Extension	*Proximally* Along and 1–2 cm lateral to the anterior crest of the tibia (see Chap. 22).
Internal Fixation (continued)	*Tibial pilon fracture* The bone is buttressed by applying a T, cloverleaf, or DC plate to the medial or anterior aspect of the distal tibia, depending on the type of fracture.
Closure	A skin suture is usually sufficient.
Hazards	Anterior tibial vessels, deep fibular nerve. Very delicate skin conditions of this area.

Fig. 24d

1   Crural fascia
2   Ankle joint (talocrural joint)
3   Medial ligament (deltoid ligament)

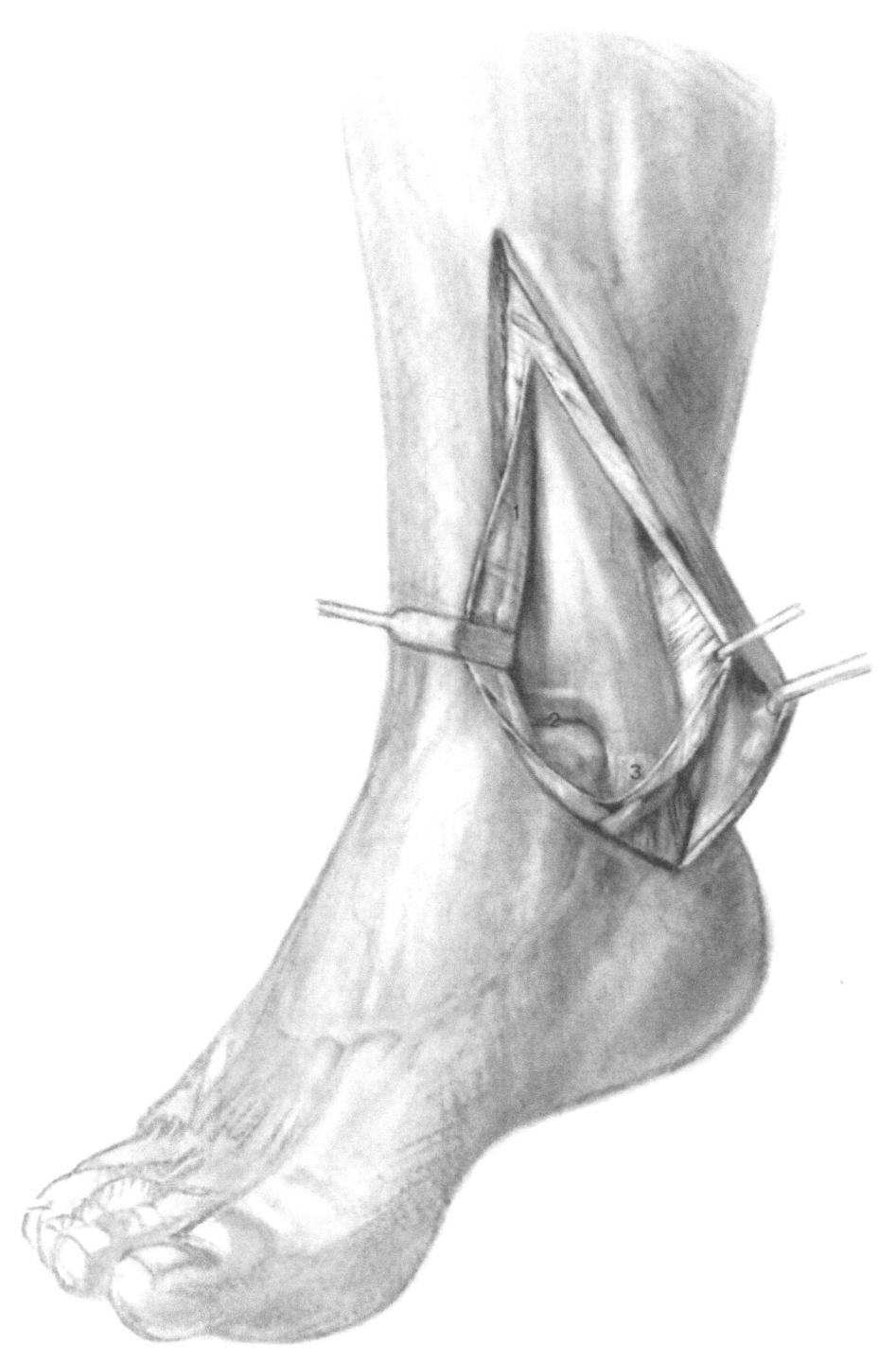

25 Lateral Malleolus

Indications   Bony or ligamentous injuries of the lateral malleolus.

Positioning   Supine with the leg in slight internal rotation; tourniquet.

Skin Incision   A straight, hockey-stick or lazy-S-shaped incision about 10 cm long is made past the palpable lateral malleolus, running either anterior or posterior to it. With a posterior incision, care is taken *not to injure* the sural nerve and small saphenous vein. With an anterior incision, the superficial peroneal nerve must be spared.

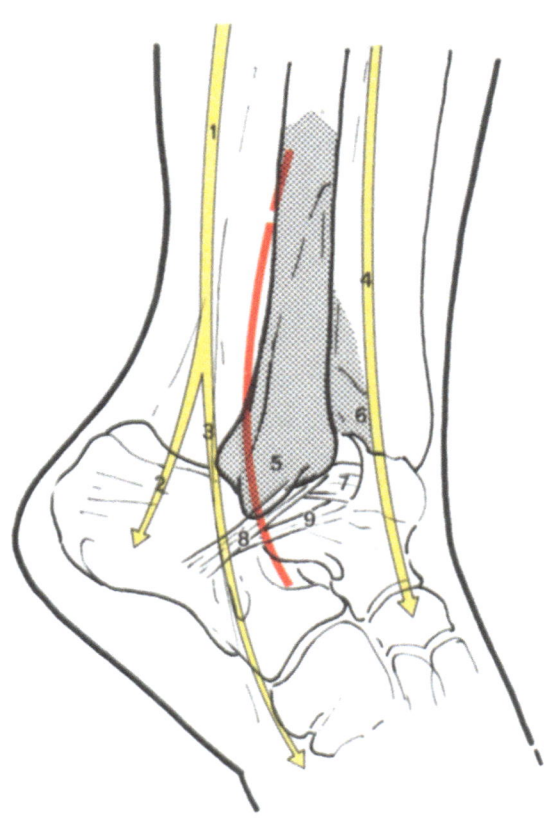

Fig. 25a

1   Peroneal nerve
2   Lateral calcaneal branches of peroneal nerve
3   Lateral dorsal cutaneous nerve
4   Intermediate dorsal cutaneous nerve
5   Lateral malleolus
6   "Tubercle of Chaput"
7   Anterior talo-fibular ligament
8   Calcaneo-fibular ligament
9   Lateral talo-calcaneal ligament

Deep Dissection   The interior of the ankle joint is inspected through the fracture line. It may be necessary to split the crural fascia between the fibula and tibia to expose the anterior syndesmosis.
Beware of injury to perforating branch of peroneal artery.

Fig. 25b

1   Peroneal muscles
2   Superficial peroneal nerve
3   Superior extensor retinaculum
4   Lateral malleolus

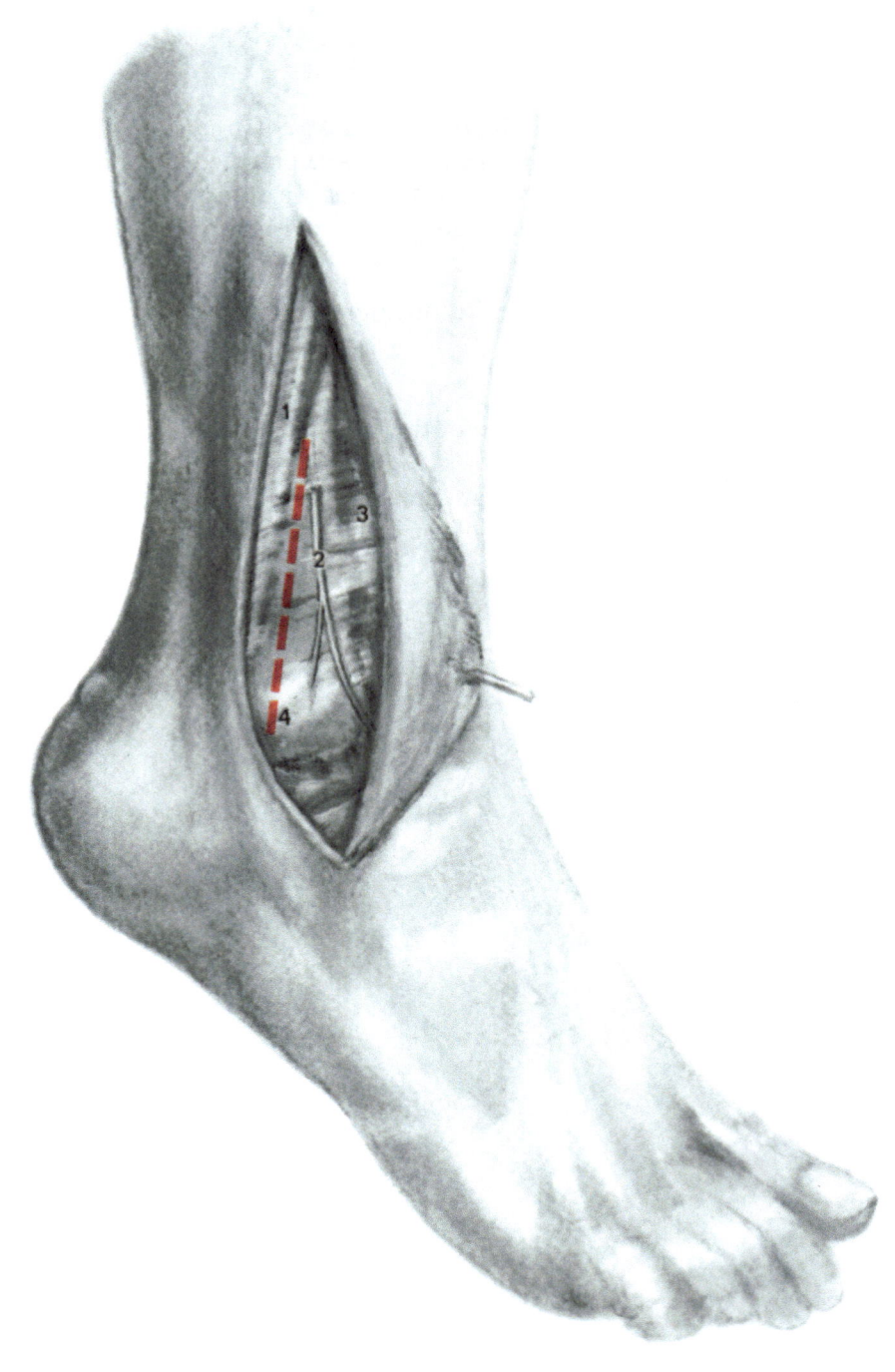

Extension
For a posterior edge fracture (VOLKMANN'S *triangle*)
The posterior edge of the tibia can be palpated and visualized by detaching and retracting the peroneal tendons. In some instances an approach from the medial or even posterior side is advisable.

Proximally
Along the fibula (beware of injury to common peroneal nerve at fibular neck).

Distally
Curved toward the head of the fifth metatarsal.

Internal Fixation
Fixation is done with lag screws or a 1/3-tubular plate applied to the postero-lateral or lateral aspect of the distal end of the fibula.

Closure
Skin suture.

Fig. 25c

1 Peroneal muscles
2 "Subcutaneous triangular area of fibula"
3 Perforating branch of peroneal artery
4 Anterior talo-fibular ligament
5 Calcaneo-fibular ligament
6 Extensor digitorum longus (peroneus tertius)

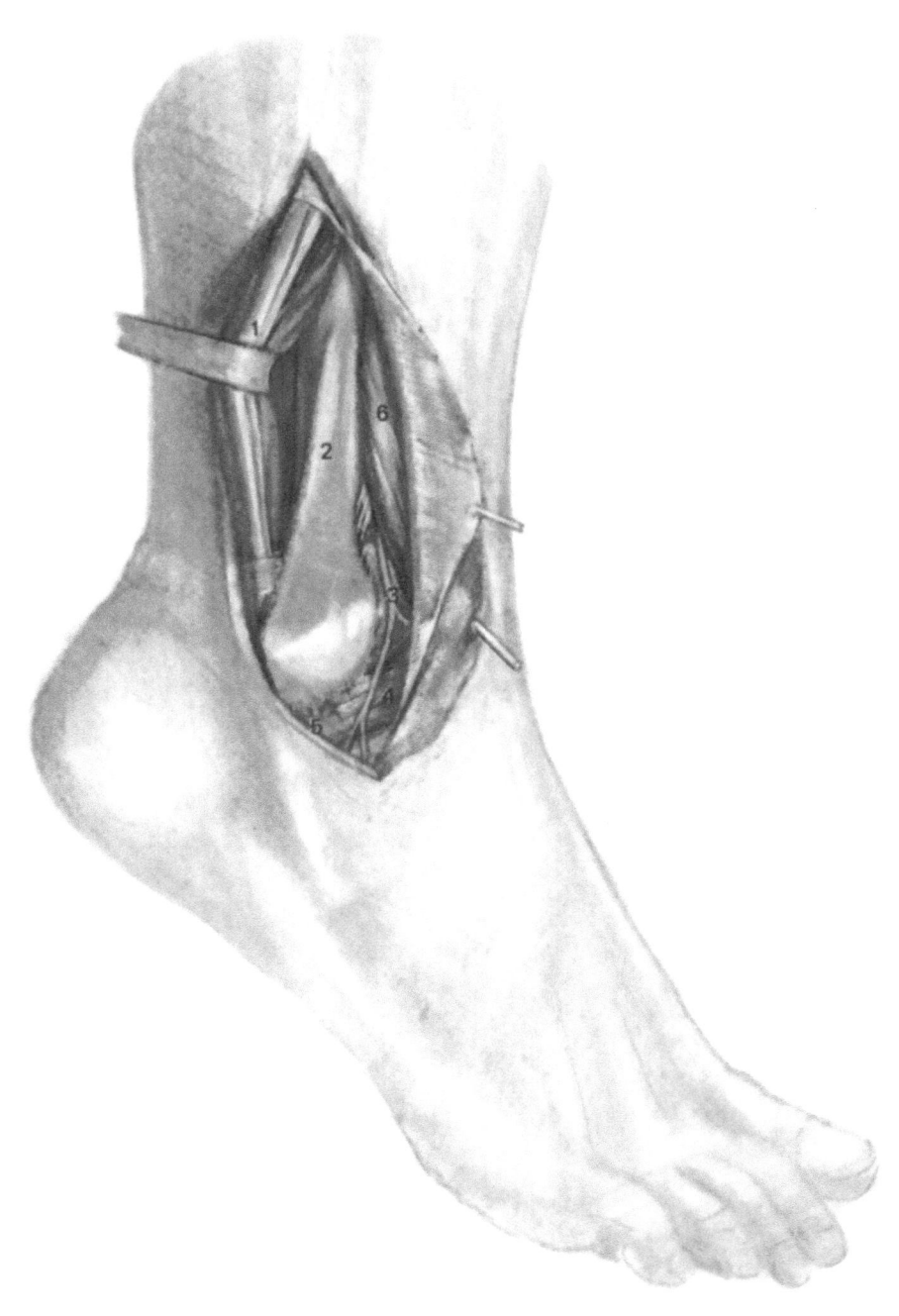

26 Calcaneus: Lateral Approach

Indications   Comminuted fractures of the calcaneus, lateral ligamentous injuries of the ankle joint.

Positioning   Supine with padding under buttock and leg rotated internally; tourniquet.

Skin Incision   Curves gently from the cleft between the *lateral malleolus* and *Achilles tendon* about 1–2 fingerwidths below the tip of the malleolus toward the head of the *fifth metatarsal.*

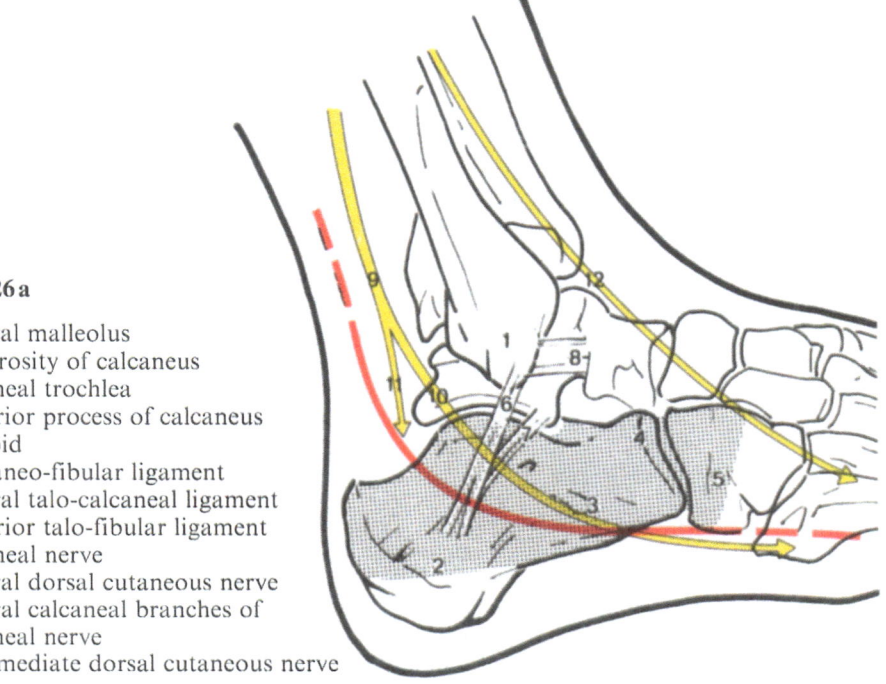

Fig. 26a

1   Lateral malleolus
2   Tuberosity of calcaneus
3   Peroneal trochlea
4   Anterior process of calcaneus
5   Cuboid
6   Calcaneo-fibular ligament
7   Lateral talo-calcaneal ligament
8   Anterior talo-fibular ligament
9   Peroneal nerve
10 Lateral dorsal cutaneous nerve
11 Lateral calcaneal branches of peroneal nerve
12 Intermediate dorsal cutaneous nerve

Deep Dissection
The superficial afferents of the small saphenous vein are ligated, sparing branches of the sural nerve that supply the heel (lateral calcaneal branches) and lateral border of the foot.

To expose the entire lateral aspect of the calcaneus, the distal peroneal retinaculum is divided and detached from the peroneal trochlea (tendon-sheath compartment of both peroneus tendons), and the tendons are displaced proximally. It may be necessary to detach the fleshy origin of extensor digitorum brevis from the calcaneus, i.e., from the floor of the tarsal canal.

Extension
Distally
Along the lateral border of the foot with exposure of the cuboid.

Proximally
Parallel to the limb axis.

Closure
The retinacula of both peroneal tendons are reattached, and the skin is closed.

Hazards
Injury to the sural nerve and its branches (lateral calcaneal branches and lateral branch of superficial peroneal nerve).

Fig. 26b

1   Superior peroneal retinaculum
2   Lateral malleolus
3   Inferior peroneal retinaculum
4   Anterior talo-fibular ligament
5   Inferior extensor retinaculum
    (cruciform ligament)
6   Extensor digitorum brevis
7   Tendon of peroneus brevis
8   Tendon of peroneus longus
9   Sural nerve and small saphenous vein

Fig. 26c

1   Sural nerve
2   Tuberosity of calcaneus
3   Peroneal trochlea and
    inferior peroneal retinaculum (cut)
4   Tendon of peroneus brevis
5   Abductor digiti minimi
6   Tarsal sinus

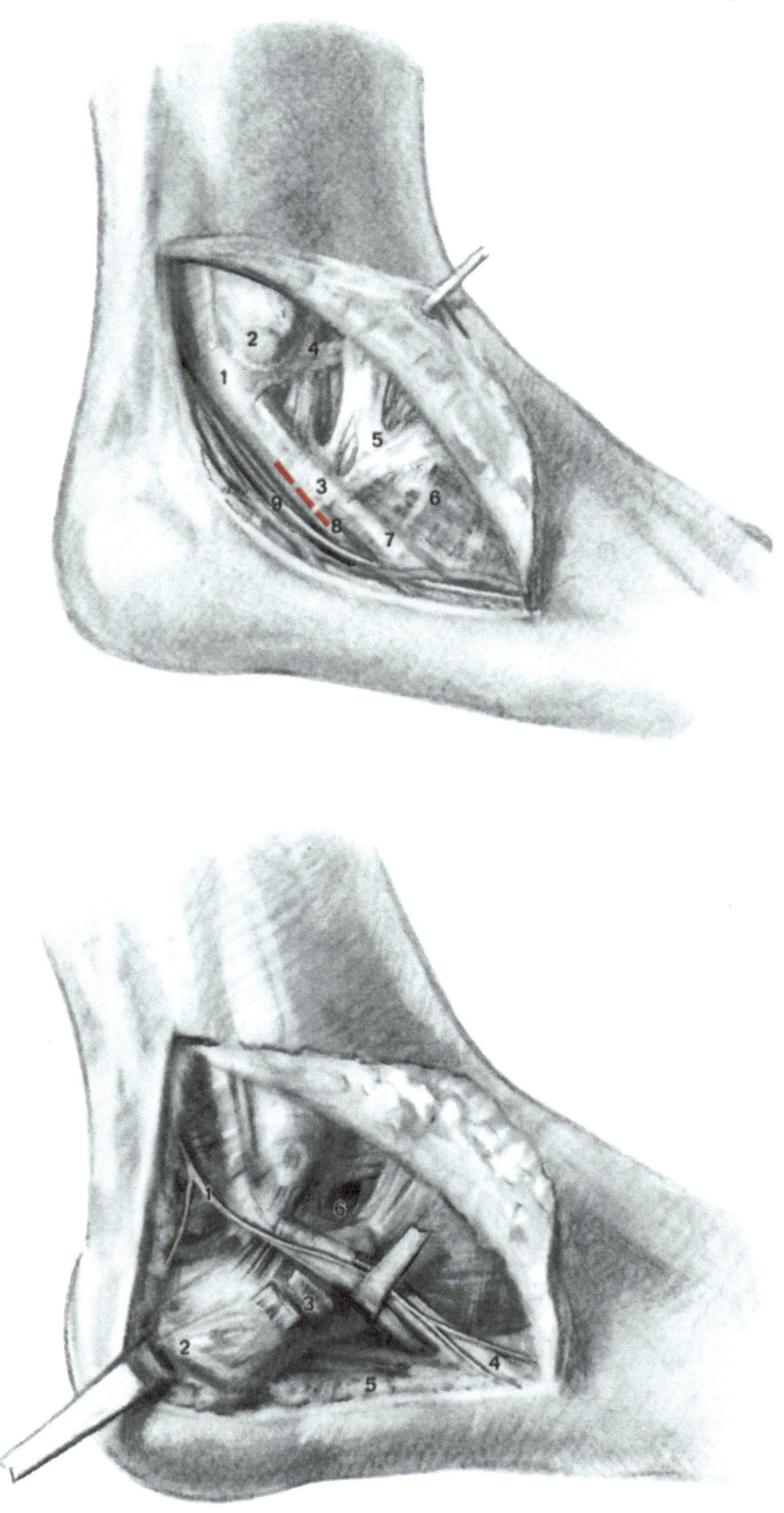

References

BRÜCKNER H, HINZE M (1980) Zugangswege in der Traumatologie. Barth, Leipzig

HENRY AK (1959) Extensile exposure. Livingstone, Edinburgh London

HONNART F (1978) Voies d'abord en chirurgie orthopédique et traumatologie. Masson, Paris New York Barcelona Mailand

LETOURNEL E, JUDET R (1981) Fractures of the acetabulum. Springer, Berlin Heidelberg New York

MÜLLER W (1982) Das Knie. Springer, Berlin Heidelberg New York

NICOLA T (1971) Atlas operativer Zugangswege in der Orthopädie. Urban and Schwarzenberg, München Berlin Wien

Subject Index

Acetabulum
    anterior column, ilio-inguinal
      109–113
    posterior column, roof
      101–107
    posterior lip 93–99
Acromio-clavicular dislocation
    3–7

Bicepstendon, distal avulsion
    49–53

Calcaneus/os calcis 181–183
Cancellous autograft
    dorsal iliac crest 77–83
    greater trochanter 115–119
Carpal-tunnel 69–73
Clavicule 3–7
Compartment syndrom
    forearm 58–67
    lower leg 159–165
Coronoid process of ulna
    37–41, 49–55

Dislocation
    acromio-clavicular 3–7
    elbow 43–55
    hip joint, posterior 93–99
    sacro-iliac joint 77–83
    shoulder, anterior 15–19
    shoulder, posterior 9–13
    tibio-talar joint 167–179

Elbowjoint
    dorsal approach 37–41
    radial approach 43–48
    volar approach 49–55

Femur
    inter-/subtrochanteric
      115–121
    neck: head prosthesis
      93–99
    neck: internal fixation
      115–121
    shaft 123–129
    supra-/transcondylar 131–137

Gerdy's tubercule 143–147
Greater tubercule of humerus
    15–19

Humerus
    capitulum 43–48
    greater tubercule 15–19
    intraarticular fracture distal
      37–41
    intraarticular fracture proxi-
      mal 15–19
    shaft, distal third 37–41
    shaft, middle third, antero-
      lateral 29–35
    shaft, middle third, dorsal
      21–27
    shaft, proximal third 15–19,
      21–27
    subcapital fracture 15–19

Iliac crest
    dorsal 77–83
    ventral 88–91, 109–113

Knee
    collateral ligaments 143–153
    cruciate ligaments 143–153
    intraarticular fractures
      143–153
    intraarticular revision
      143–153

Ligamentons,
    repair/reconstruction
    elbow 37–48
    knee, collaterals 143–153
    knee, cruciates 143–153
    malleoli, lateral 181–183
    malleoli, medial 167–169

Malleoli
    lateral 167–169
    ligaments lateral repair/recon-
      struction 181–183
    ligaments medial 167–169
    medial 175–179
    Volkmann's Triangle 178–179
Monteggia fracture 43–48,
    63–67

Olecranon 37–41
Osteotomy
    acromion 18–19
    anterior superior iliac spine
      88–91, 110–111

Gerdy's tubercule 145–147
greater trochanter 104–105
lateral humeral epicondyle
    44–45
olecranon 38–39
Pes anserinus 151–153
tibial tuberosity 136–137

Patella 139–141
Pelvic ring
    pubic rami 85–91,
      109–113
    sacral ala 82
    sacro-iliac joint 77–83
    symphysis pubis 85–87
Pilon tibial
    fibular approach 175–179
    tibial approach 167–173
Pubic rami 85–91, 109–113

Radius
    distal end, intraarticular
      69–73
    head 43–53
    shaft, dorsal approach
      57–61
    shaft, volar approach
      43–55

Sacro-iliac joint 77–83
Shoulderjoint
    anterior dislocation 15–19
    fracture of glenoid 9–13
    posterior dislocation
      9–13
Symphysis pubis 85–87

Tibia
    distal end, intraarticular
      167–173
    medullary nailing 139–141,
      156
    plateau 143–153
    shaft, anterior approach
      155–157
    shaft, medio-dorsal approach
      159–165
    tuberosity, osteotomy
      136–137

Ulnarshaft 63–67

Manual of Internal Fixation

Techniques Recommended by the AO Group

By **M. E. Müller, M. Allgöwer, R. Schneider, H. Willenegger**
In collaboration with numerous experts
Translated from the German by J. Schatzker

2nd, expanded and revised edition. 1979. 345 figures in
color, 2 Templates for Preoperative Planning.
X, 409 pages
ISBN 3-540-09227-7

C. F. Brunner, B. G. Weber

Special Techniques in Internal Fixation

Translated from the German by T. C. Telger
1982. 91 figures. X, 198 pages
ISBN 3-540-11056-9

U. Heim, K. M. Pfeiffer

Small Fragment Set Manual

Technique Recommended by the ASIF Group
Translated from the German by R. L. Batten and
K. M. Pfeiffer

2nd expanded and revised edition. 1982. 215 figures in more
than 500 separate illustrations. IX, 396 pages
ISBN 3-540-11143-3

F. Séquin, R. Texhammar

AO/ASIF Instrumentation

Manual of Use and Care

Introduction and Scientific Aspects by H. Willenegger
Translated from the German by T. C. Telger

1981. Approx. 1300 figures, 17 separate Checklists.
XVI, 306 pages
ISBN 3-540-10337-6

Springer-Verlag
Berlin
Heidelberg
New York
Tokyo

Springer AV Instruction Programme

Films/Videocassettes:

Theoretical and practical bases of internal fixation, results of experimental research:

Internal Fixation – Basic Principles and Modern Means
The Biomechanics of Internal Fixation
The Ligaments of the Knee Joint. Pathophysiology

Internal fixation of fractures and reconstructive bone surgery:

Interal Fixation of Forearm Fractures
Internal Fixation of Noninfected Diaphyseal Pseudarthroses
Internal Fixation of Malleolar Fractures
Internal Fixation of Patella Fractures
Medullary Nailing
Internal Fixation of the Distal End of the Humerus
Internal Fixation of Mandibular Fractures
Corrective Osteotomy of the Distal Tibia
Internal Fixation of Tibial Head Fractures (available in German only)

Joint replacement:

Total Hip Prostheses (3 parts)
Part 1: Instruments, Operation on Model
Part 2: Operative Technique
Part 3: Complications. Special Cases
Elbow-Arthroplasty with the New GSB-Prosthesis
Total Wrist Joint Replacement

Replantation surgery:
Microsurgery for Accidents

Slide Series:

ASIF-Technique for Internal Fixation of Fractures
Manual of Internal Fixation
Small Fragment Set Manual
Internal Fixation of Patella and Malleolar Fractures
Total Hip Prosteses
Operation on Model and in vivo, Complications and Special Cases

● Please ask for information material
● Order from:
Springer-Verlag Heidelberger Platz 3, D-1000 Berlin 33, or Springer-Verlag New York Inc., 175 Fifth Avenue, New York, Ny 10010, USA

Springer-Verlag
Berlin
Heidelberg
New York
Tokyo